Exploring Science

Human Biology and Evolution

Exploring Science

Human Biology and Evolution

Lisa Sita

Thomson Learning
New York

A FRIEDMAN GROUP BOOK

First published in the
United States in 1995 by
Thomson Learning
New York, NY

Library of Congress Cataloging-in-Publication Data
Sita, Lisa, date
Human biology and evolution / Lisa Sita
p. cm.—(Exploring science)
Includes bibliographical references and index.
ISBN 1-56847-270-6
1. Human biology—Juvenile literature. 2. Human evolution—Juvenile literature. [1. Human biology. 2. Evolution.]
I. Title. II. Series: Exploring science (New York, N.Y.)
QP37.S58 1995
573—dc20

94-46316
CIP
AC

EXPLORING SCIENCE: HUMAN BIOLOGY AND EVOLUTION
was prepared and produced by
Michael Friedman Publishing Group, Inc.
15 West 26th Street
New York, New York 10010

Editor: Nathaniel Marunas
Art Director/ Designer: Jeff Batzli
Photography Director: Christopher Bain
Photography Researchers: Christopher Bain and Susan Mettler
Illustrations: George Gilliland, Laura Pardi Duprey, and Siena Artworks Ltd.

Front Cover Illustration: George Gilliland
Back Cover Illustrations: Siena Artworks (left); Laura Pardi Duprey (right)

Color separations by Benday Scancolour Co. Ltd.
Printed in China.

Dedication

For my parents, Margaret and Anthony Sita.

Acknowledgments

The author offers her sincere thanks to her friends and colleagues in the Education Department of the American Museum of Natural History, especially Karen Lund, Brad Burnham, Karen Kane, Ann Prewitt, and Marcia White; Nathaniel Marunas and Jeff Batzli of Michael Friedman Publishing; Kelly Matthews; Thomson Learning; and friends and family members, especially Joey, Stephanie, Tonie-Ann, and Manny, for their patience and support during the writing of this book.

Table of Contents

INTRODUCTION

LIFE'S BEGINNINGS

THESE CHILDREN ARE ENJOYING A GAME OF TAG IN HOUSTON, TEXAS. ALTHOUGH IT SEEMS EASY ENOUGH, RUNNING IS A COMPLICATED ACTIVITY THAT WOULD NOT BE POSSIBLE WITHOUT THE BRAIN SENDING AND RECEIVING VAST AMOUNTS OF INFORMATION.

THE HUMAN SPECIES

The human body is a fascinating life form that is specially equipped to take care of all our needs. No machine, including the most advanced computer, can match it in complexity. The human body's millions of parts work together to let us move, sleep, eat, think, and do all the things we do every day. Most of the time, we are not even aware of all the work our bodies do for us. Let's explore, for example, what happens when we play a game of tag.

Before we run to tag our opponent, our brains send messages through the nerves that tell our muscles to move. Using our eyes and ears, we can tell where our opponent is and then move in that direction. As we run, our brain sends instructions through the nerves to our muscles in order to keep us balanced and upright. Our lungs breathe faster and more deeply to bring more oxygen into our bodies. Our hearts beat faster to carry that oxygen in our blood to our muscles. Our sweat glands produce salt and water to cool our bodies.

The Study of Life

The word **biology** means "the study of life." **Human biology** is the study of the human body and how it works.

When we tag our opponent, the nerve endings in the skin of our fingertips allow us to feel the contact.

We may rest after the game, but our bodies never rest completely. Even when we are asleep, our hearts pump blood, our lungs breathe air, our muscles move, and our brains keep sending messages to all our body parts, telling them what to do.

The human brain, so special and complex, is one of the things that makes us unique, enabling us to think, reason, and experience emotion. Our advanced brains (which enable us to store and analyze mountains of data) and specialized bodies have allowed us to build cities and invent machines, to create art and written languages, to discover cures for diseases, and to do many other amazing things. But, just as a baby takes many years to grow and develop into an adult, the human species has taken a long time (millions of years) to become what it is today.

A VOLCANO SPEWS FIRE, GAS, AND ROCK IN HAWAII. THROUGHOUT THE EARTH'S HISTORY, NATURAL OCCURRENCES SUCH AS THIS HAVE CHANGED THE WAY THE SURFACE OF THE PLANET LOOKS.

Our Busy Bodies

Every day our bodies work to keep us fit and healthy. When we are sick, our bodies work to fight the illness. Most of the time, we don't even think about all the work our bodies do.

You will need: a pencil and a piece of paper.

Take an hour out of your day to pay attention to your body's activities. Try to write down *everything* you do in that hour. Be sure to include such things as eating, walking, and talking to a friend on the phone. Later, go over your list and think about all the parts of your body that were working together to help you do those things. You will be surprised at the results.

DARWIN'S THEORY

When we examine the evidence from the distant past, we find that animals and plants have not always looked like they do today. A few hundred thousand years ago, for example, saber-toothed tigers and woolly mammoths walked the Earth. These animals are no longer around, but their modern relatives, the tigers and the elephants, are. Even the planet Earth itself has changed over the 4.5 billion years since it first appeared in the universe: continents have shifted; mountains have risen out of the sea; and the crust has been constantly reshaped by earthquakes, volcanoes, and the steady movement of **glaciers,** which are huge bodies of ice that slowly cross the Earth.

How and why did life on Earth change? To answer this question, we must look to a man named **Charles Darwin** and his **theory of evolution**. A **theory** is an idea about how something works that has yet to be proven; sometimes the evidence supporting such an idea is so strong that the scientific world accepts it as fact. This is true of Darwin's theory of evolution. In Darwin's time and before, other great thinkers had come up with similar theories, but Darwin was the first person to put these ideas together and then add something of his own, in a book published in 1859 called *On the Origin of Species*.

Charles Darwin was born in England in the year 1809. Although he had studied first to be a doctor and then a clergyman, Darwin was always more interested in studying the world of nature than anything else. In 1831, when the British ship *Beagle* set sail on a scientific expedition around the world, Darwin was invited to go along as the ship's naturalist. For five years he traveled aboard the *Beagle*, studying the plants, animals, and rocks of many remote places.

THE NATURALIST CHARLES DARWIN (1809–1882).

Changing Species

Although dogs and wolves are different species today, they share a common ancestor. The process of evolution is responsible for their differences, which arose over time.

THE *BEAGLE* CARRIED DARWIN TO THE SCENE OF MANY OF HIS MOST IMPORTANT DISCOVERIES.

As he explored, Darwin began to realize something about some of the different kinds, or **species**, of plants and animals he found. He realized that some of these species were alike in many ways, although they lived on different islands. Darwin began to develop a theory about why this was so. He proposed that all forms of life evolved in stages through long periods of time. "Change through time" is what evolution means. According to Darwin, evolution happens in this way: every species can produce babies, and every baby within that species will be a little bit different in some way from every other member of the

THE BROWN PELICAN IS ONE OF THE MANY SPECIES DARWIN STUDIED IN THE GALAPAGOS ISLANDS, ECUADOR. DARWIN'S THEORY OF EVOLUTION WAS GREATLY INFLUENCED BY THE DISCOVERIES HE MADE IN THESE ISLANDS.

species. In some cases, individuals will be born with certain traits that will help them to survive and to reproduce. For instance, an individual might be born with a trait that enables it to exploit its food resources more fully.

As the individual grows up, it may face certain problems—there may not be enough available food or there may be a population of predators in the vicinity. Those individuals who have the necessary survival traits will live, while the other members of the species will not. Darwin called this **natural selection**. The survivors then have offspring of their own and pass those survival traits on to the next generation. When a new generation picks up a change that will help it survive, it is called **adaptation**.

Through long periods of time, as these traits are passed from generation to generation, the entire species may change and eventually become a new species, especially if changes in the environment also occur. For example, a mountain range may build up in an area where a species lives. This range may separate the population; because the two groups no longer reproduce with each other, they may begin to evolve into two different species. This process is called **speciation**.

Like other forms of life, humans have evolved through various species to where we are today. The story of how we became what we are is an adventure that began nearly four million years ago in the

Natural Selection and Adaptation

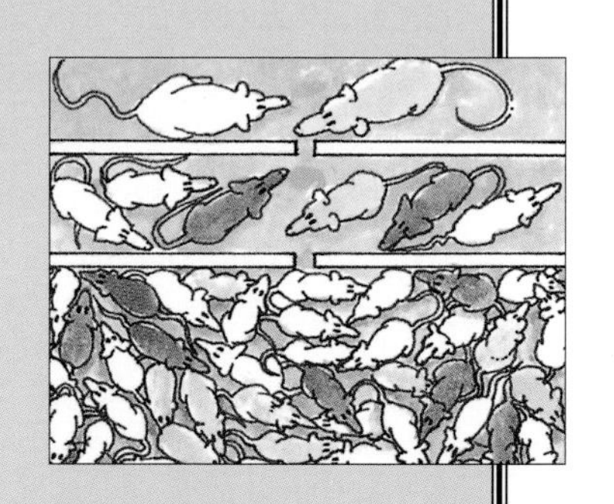

Every species produces babies, and each baby is a little bit different from the other members of the species.

Not all individuals will survive. Some may be eaten by predators, for example, or some may not be able to get enough food.

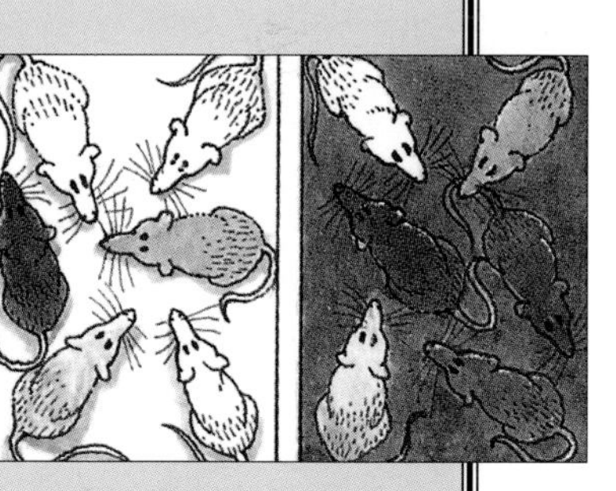

Because the individuals are different, some will have traits that may help them to survive better than other individuals. For example, some may have patterns and colors on their bodies that blend better into the environment, which makes them less noticeable to predators. In these drawings, the mice with darker coats blend better into the dark soil of their surroundings than do the mice with lighter coats.

The darker mice are more likely to live and produce offspring. If nothing changes in the environment, the offspring will also have dark coats. The offspring will then pass this trait on to their own offspring. Eventually the majority of the mice in that area will have dark coats.

grasslands of southern and eastern Africa. But in order to truly understand where we came from, we must first journey back 3.5 billion years to when the first stirrings of life appeared on our planet.

BILLIONS OF YEARS AGO, RAINSTORMS AND LIGHTNING CRASHED THROUGH THE EARTH'S ATMOSPHERE, FORMING VAST, EMPTY OCEANS.

WHEN LIFE FIRST BEGAN

Imagine a planet where hot, molten rock bursts upward out of the ground and sprays across the planet's surface, sending out deadly gases like methane and ammonia. Rainstorms and lightning crash through this atmosphere to form vast oceans empty of life. Imagine a planet with no oxygen, no animals, and no plants. That was what Earth used to be like. Suddenly, some kind of energy, either lightning or the sun's radiation, fuses together some **molecules** (tiny particles of matter) from the water and the atmosphere. (In this case, the molecules were particles of gas and water.) These special chains of molecules twist together, and then began absorbing other molecules to nourish them-

Compare the age of the Earth to a twenty-four-hour period, with each hour representing 200 million years. With the Earth first appearing at midnight, our earliest human ancestors would have appeared at about one minute before midnight the following day.

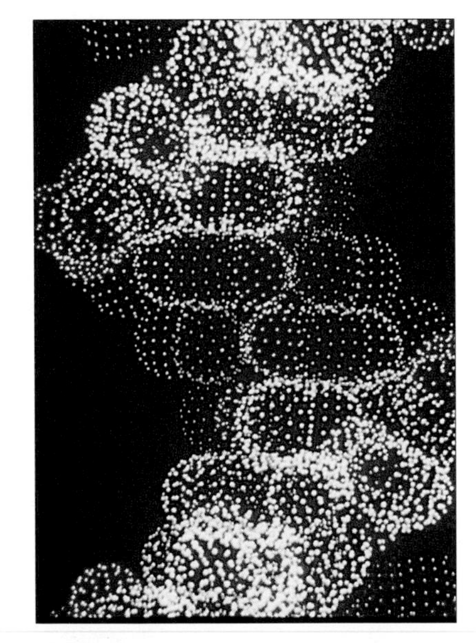

THE FIRST SIGNS OF LIFE ON EARTH APPEARED WHEN MOLECULES FUSED TOGETHER AND FOUND A WAY TO REPRODUCE THEMSELVES. SCIENTISTS OFTEN CONSTRUCT MODELS OF WHAT MOLECULES LOOK LIKE, AS SHOWN HERE.

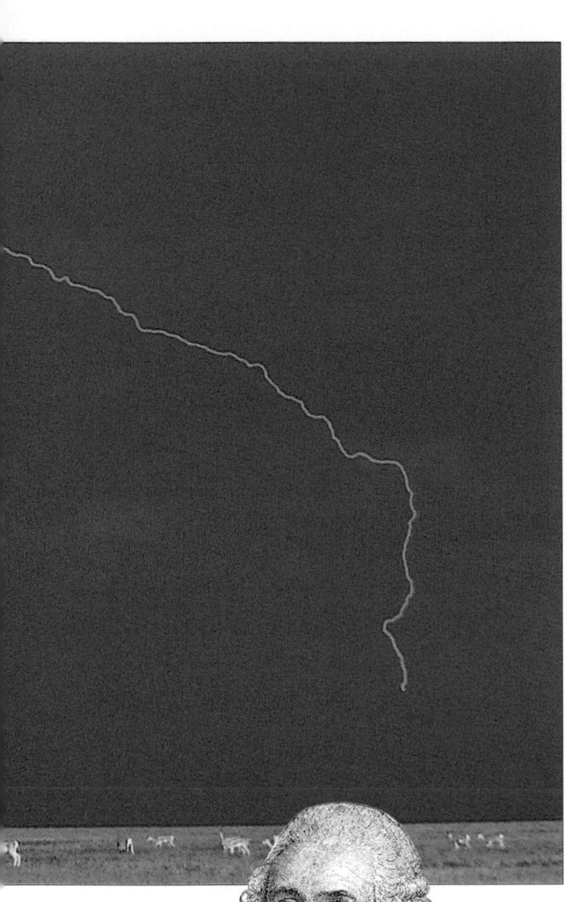

An Important Contribution

Carolus Linnaeus, a Swedish naturalist and physician of the eighteenth century, developed a system for categorizing plants and animals that became the standard for the whole world. He classified organisms by giving them two names, a **genus** name and a **species** name. For instance, modern humans belong to the genus *Homo* and the species *Homo sapiens*.

Growing Bread Mold

Bacteria was the earliest form of life on Earth. In time, bacteria grew to form other kinds of life—primitive plants like **algae** and **fungus**. These types of plants grew quickly, spreading rapidly over the earth.

To see how quickly fungus can grow and spread, try growing some mold of your own.

You will need: a piece of bread, a few drops of water, and a clear plastic or glass jar with a lid.

1. Place a few drops of water on a piece of bread and put the bread in a jar.

2. Poke some holes in the lid of the jar and place it in a warm, dark place, like a closet or cabinet.

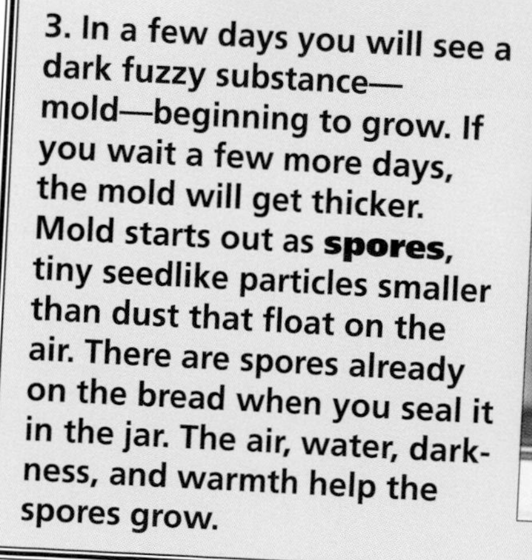

3. In a few days you will see a dark fuzzy substance—mold—beginning to grow. If you wait a few more days, the mold will get thicker. Mold starts out as **spores**, tiny seedlike particles smaller than dust that float on the air. There are spores already on the bread when you seal it in the jar. The air, water, darkness, and warmth help the spores grow.

selves. Over the next billion years or so these molecules find ways of reproducing.

That is how life began on our planet. These earliest forms of life were **bacteria**—one-celled organisms so small that they can only be seen with a microscope. During the next two billion years, some bacteria were able to use the sun's energy to produce oxygen. The oxygen formed a protective blanket around the Earth. This was the Earth's new **atmosphere**, and it allowed the future species of plants and animals to begin evolving.

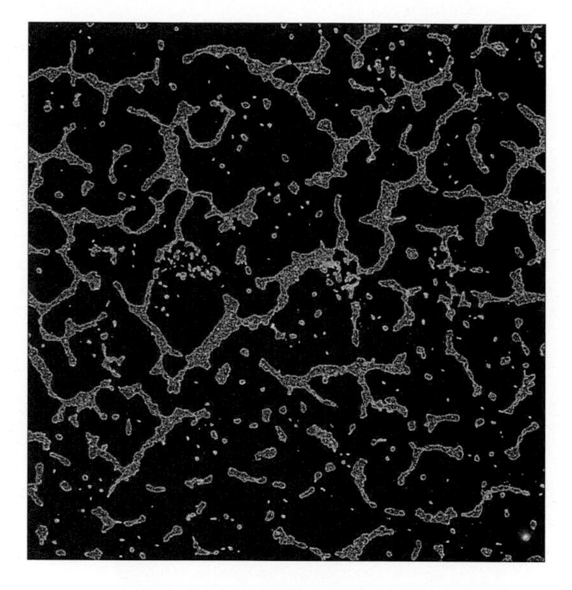

THIS IS WHAT BACTERIA LOOKS LIKE—MAGNIFIED TWO HUNDRED TIMES.

Humans Are Latecomers

Earth's history is divided into many periods of time, the names of most of which appear below. As you can see, humans did not appear on the scene until very late in the game.

The Precambrian Era—more than 570 million years ago:
- **4.5 billion years ago—Earth was formed.**
- **3.5 billion years ago—Bacteria appeared in Earth's waters (the first signs of life on Earth).**
- **1.3 billion years ago—Algae appeared in Earth's waters.**
- **670 million years ago—Multicelled animals appeared in Earth's waters.**

The Paleozoic Era—570 million years ago to 245 million years ago:
- **570–510 million years ago (The Cambrian Period)—Hard-shelled animals without backbones appeared in Earth's waters.**
- **510–439 million years ago (The Ordovician Period)—Fish and shellfish appeared.**
- **439–408 million years ago (The Silurian Period)—Early plant life appeared on land.**
- **408–362 million years ago (The Devonian Period)—First land animals (insects, spiders, scorpions, and amphibians) appeared. Forests began to grow.**

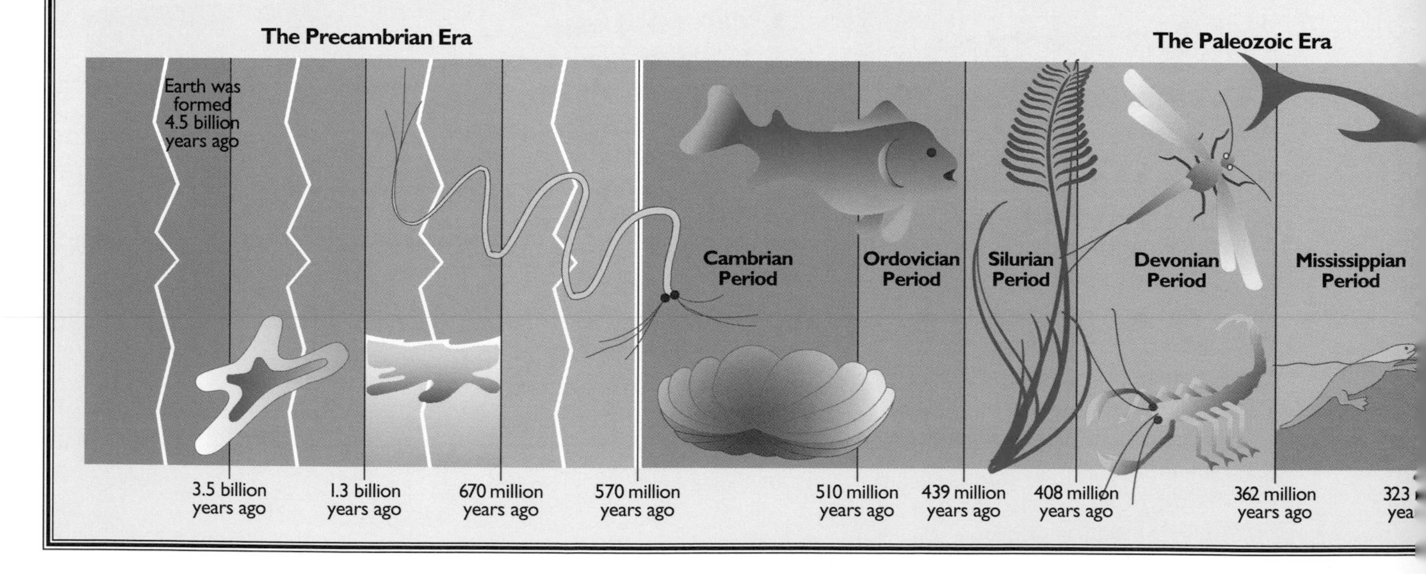

Above, left: Insects were among the first land animals to appear on Earth, 408 to 362 million years ago. Above, right: The ancestors of animals like this gray reef shark first appeared in the Earth's waters 323 to 290 million years ago.

362–323 million years ago **(The Mississippian Period)**—First reptiles and amphibians appeared.

323–290 million years ago **(The Pennsylvanian Period)**—Reptiles and amphibians continued to evolve. Sharks appeared.

290–245 million years ago **(The Permian Period)**—Mammal-like reptiles appeared on land and large coral reefs appeared in the water. By 245 million years ago, nearly 95 percent of all species had become extinct.

The Mesozoic Era—245 million years ago to 65 million years ago:

245–208 million years ago **(The Triassic Period)**—First dinosaurs and first mammals appeared.

208–146 million years ago **(The Jurassic Period)**—Age of the dinosaurs. First birds appeared.

146–65 million years ago **(The Cretaceous Period)**—Dinosaurs still important. First flowering plants appeared. By 65 million years ago, dinosaurs had become extinct.

The Cenozoic Era—65 million years ago to the present:

65–1.8 million years ago **(The Tertiary Period)**—Mammals continued to evolve. Early human ancestors appeared 3.5 million years ago.

1.8 million years ago–present **(The Quaternary Period)**—Modern humans were found throughout the world by 40,000 years ago.

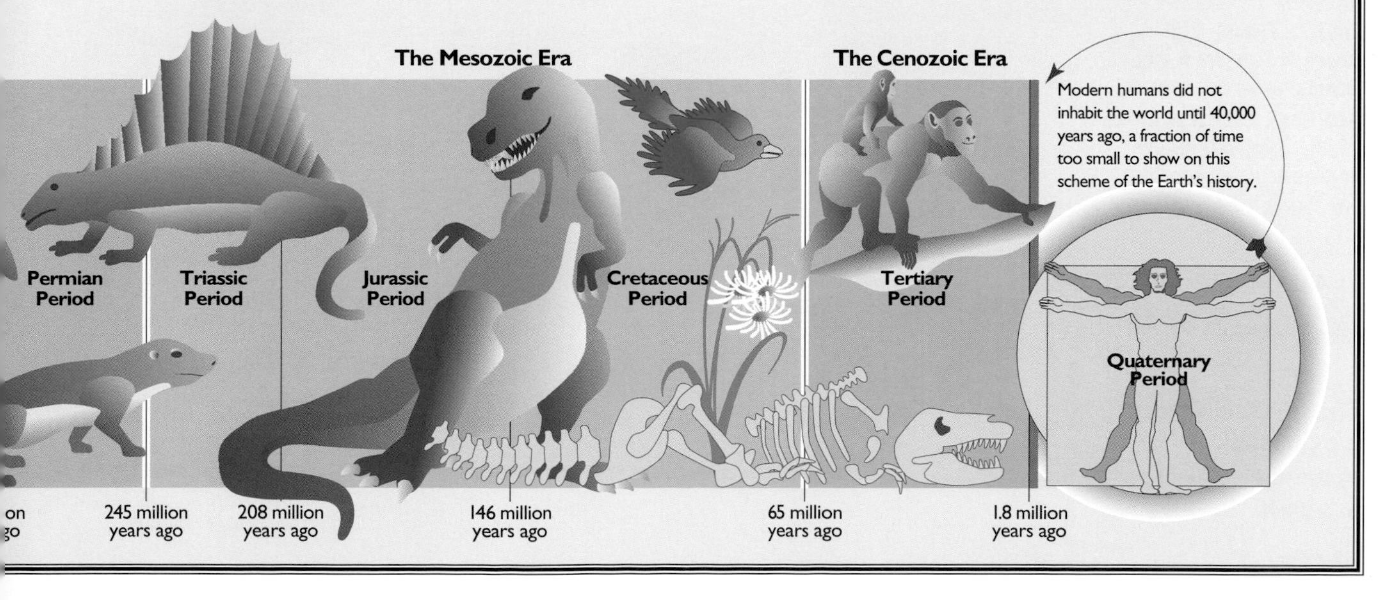

CHAPTER ONE

OUR AMAZING BODIES

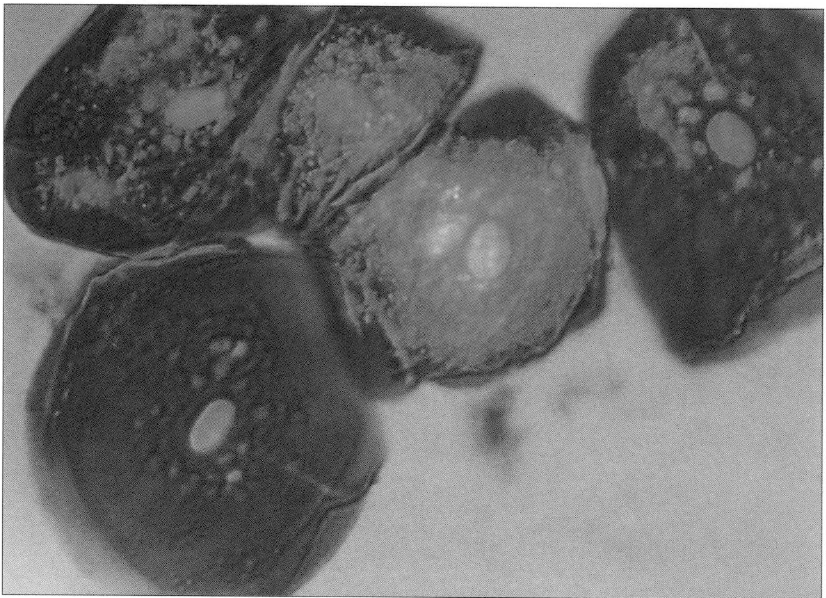

THE BUILDING BLOCKS OF OUR BODIES

We humans are many things: we are animals; we are **vertebrates**, which means we have backbones; and we are **mammals,** which means we have hair and are warm-blooded, oxygen-breathing vertebrates whose females produce live babies and milk.

Humans are complex creatures, but we share something very important with all living things on the Earth: **cells**. Some living things are made up of a single cell, while other living things—like you—are made up of many **cells**. If you hold your hand

ABOVE: CELLS ARE THE BUILDING BLOCKS OF OUR BODIES. ALL LIVING THINGS ON EARTH ARE MADE UP OF CELLS. RIGHT: THERE ARE MANY PARTS TO A CELL. LABELLED HERE ARE THE NUCLEUS, WHICH HOUSES DNA, AND THE CELL MEMBRANE, WHICH IS THE "SKIN" OF THE CELL.

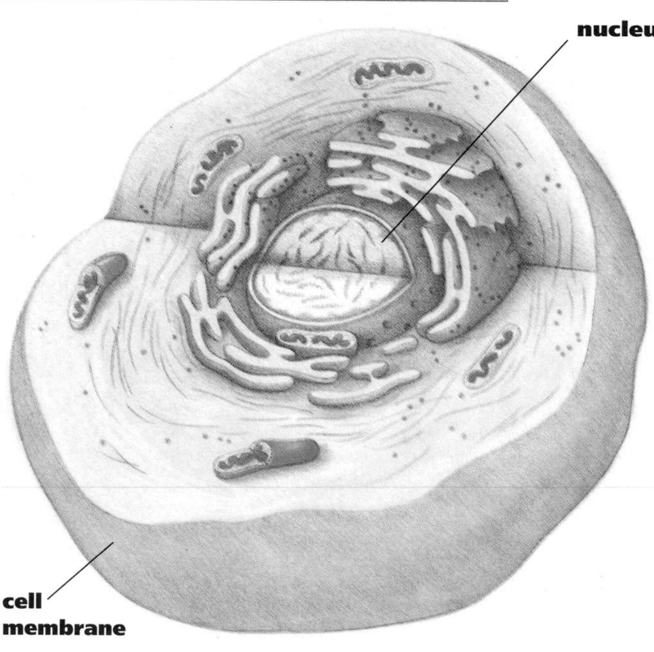

Cells Galore

There are 100 trillion (100,000,000,000,000) cells in the average adult human's body.

What Makes Something Alive?

All forms of life, from plants to elephants, have four things in common:

1. All have **carbon, hydrogen, nitrogen**, and **oxygen** in their bodies.

2. All are made up of cells.

3. All must do the same kinds of things, such as eat and breathe, to survive.

4. All depend on parts of their environment, such as air and water, to survive.

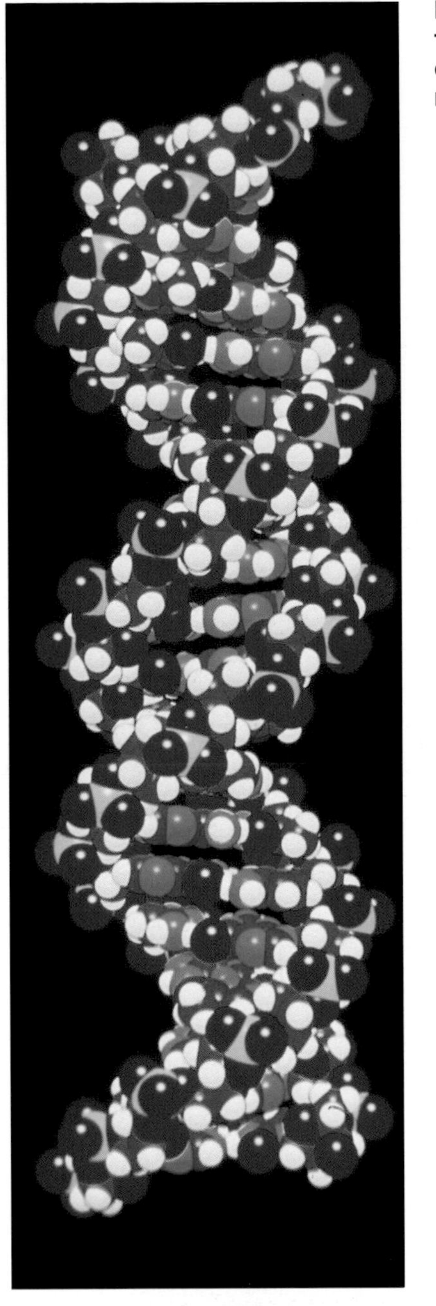

DNA IS AN ORGANIC MOLECULE THAT CONTROLS THE DEVELOPMENT OF EVERY LIVING THING ON EARTH, FROM BACTERIA TO WHALES.

very close to your face, you will see the lines and patterns of your skin. But if you look at your hand under a strong microscope, you will see that your skin is really made up of millions of tiny cells.

Each cell is a busy little world in itself, carrying out various functions necessary for its survival. Each cell takes in and digests food, gets rid of waste, heals wounds, and does just about everything else the body does as a whole. A cell has lots of parts, called **organelles**, that help it to do these things. There is also a control center, called the **nucleus**,

The Body's Little Soldiers

Floating inside the blood are special cells that help the body to fight disease and to eliminate dead or damaged cells. These special cells are the **white blood cells**, or **leukocytes**. There are several different kinds of leukocytes, each of which carries out a different aspect of the disease-fighting and cell-elimination processes. When bacteria or other dangerous organisms enter the body, the body sends a rush of blood to the infected area. When this happens, certain white blood cells identify the dangerous material, and then other white blood cells (called **macrophages**) eliminate that material (in fact, some leukocytes carry out more than one task). In this illustration, old red blood cells are being engulfed and digested by macrophages.

UNDER A MICROSCOPE, THESE WHITE BLOOD CELLS APPEAR PURPLE.

that makes sure everything is done correctly. The cells of living organisms contain **deoxyribonucleic acid (DNA)**, located inside the nucleus. DNA is made up of four chemicals and is shaped like a twisted ladder. DNA tells the cells how to make proteins, which are used to build the tissues and organs of the body. The DNA also tells the cells what color to make your hair, eyes, and skin; how tall you will be when you are fully grown; what shape your nose will be; and everything else about you. Your DNA plays a huge role in making you what you are.

Discovering Your Own Special Pattern

Your fingerprints are your own personal sign. Of all the billions of people in the world, no two of them have the same patterns of lines on the skin of their fingertips. These patterns are formed before you are born and remain the same throughout your life. By following these instructions you can find your own pattern.

You will need: two sheets of white paper, a pencil, clear tape, and a magnifying glass.

1. Rub a pencil across a sheet of white paper until a layer of graphite forms.

2. Rub one of your fingers across the graphite.

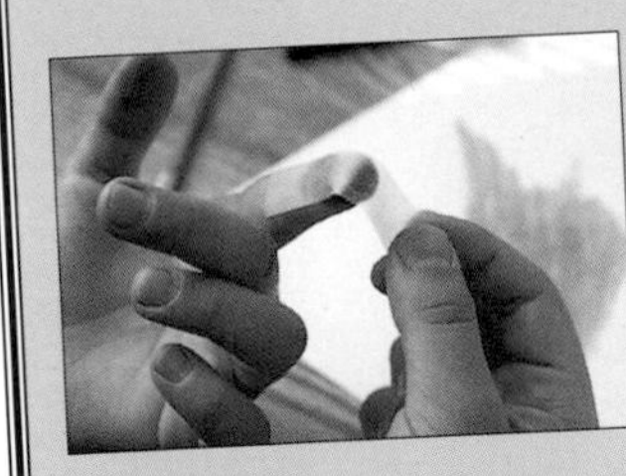

3. Put a piece of clear tape on the smudged fingertip and then remove it so that the tape lifts the graphite from your finger.

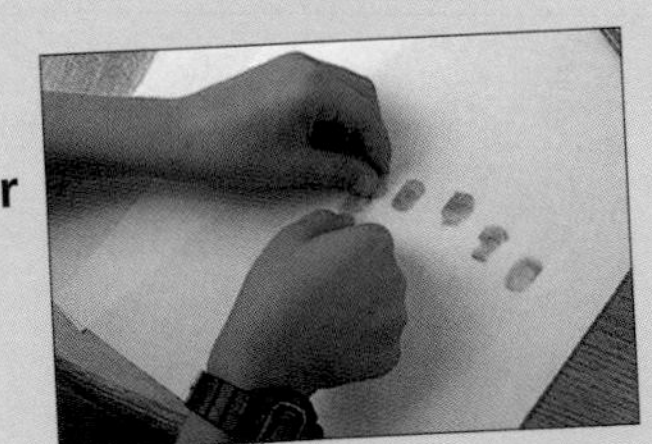

4. Place the tape on another sheet of paper.

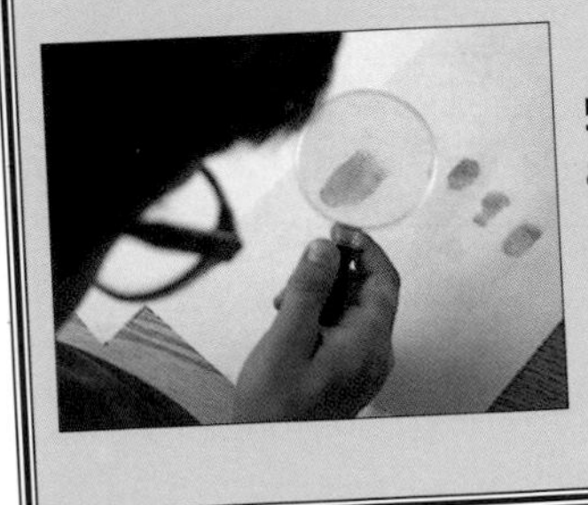

5. Do the same with your other fingers and then examine the prints with the magnifying glass. When you examine them, you will find that the prints on all the fingers are the same. They are your own special pattern.

INCREDIBLE ORGANS

Similar cells clustered together make tissue. There are four different kinds of tissues in your body: **nerve tissue** carries messages to the brain and from the brain back to your body; **muscle tissue** is elastic, like a rubber band, and allows you to move; **epithelial tissue** forms a covering for your entire body (your outer layer of skin) and also forms linings for some of your organs (for instance, your stomach and mouth); and **connective tissue** holds other tissues together.

Different kinds of tissues working together make organs. An organ is a part of the body that performs some special function. Some of the organs found in your body include the brain, heart, and lungs.

Our Incredible Skin

The largest organ of the body, and the one that grows the fastest, is the skin.

THE BRAIN IS MADE UP OF THREE PARTS: THE CEREBRUM, THE CEREBELLUM, AND THE BRAIN STEM. THE BRAIN STEM SITS ON TOP OF THE SPINAL CORD, WHICH IS THE MAIN NEURAL PATHWAY CONNECTING THE BRAIN WITH THE REST OF THE BODY.

cerebrum
brain stem
cerebellum
spinal cord

The Brain

The brain sits at the top of your spinal cord. It sends and receives messages along a webbed roadway of nerves that extend to all parts of your body.

The brain is made up of three parts: the **cerebrum**, the **cerebellum**, and the **brain stem**. The cerebrum is the largest part of your brain. It is the center of learning, thinking, and creativity, and it controls your voluntary movements and some of your emotions. The cerebellum is responsible for keeping you balanced when you walk or run and for making your muscles move the way you want them to. The brain stem takes care of most of the things your body does automatically, such as breathe, digest, and maintain the heartbeat.

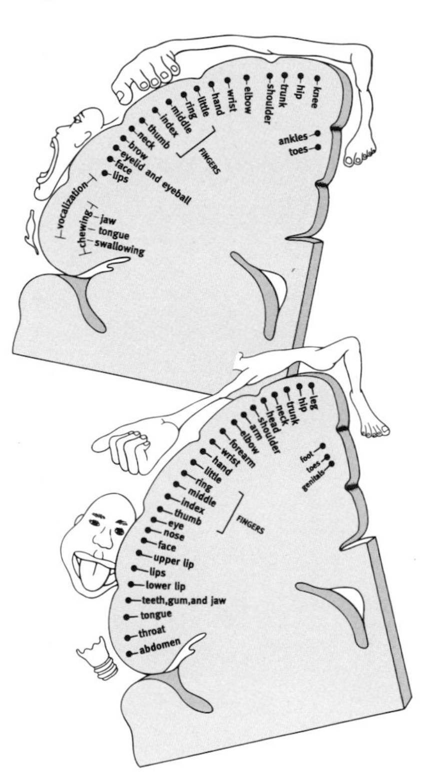

MOST OF THE BRAIN'S ANALYTICAL THOUGHT PROCESSES ARE CARRIED OUT BY DIFFERENT REGIONS OF THE CEREBRUM, WHILE THE BODY'S MOVEMENTS ARE CONTROLLED MAINLY BY THE CEREBELLUM. THESE CROSS-SECTIONS SHOW WHICH AREAS OF THE CEREBELLUM CONTROL WHICH PARTS OF THE BODY. DISCOVERIES IN BRAIN ANATOMY HAVE REVEALED THAT THE CEREBELLUM, WHICH WAS THOUGHT TO BE SOLELY INVOLVED IN MOTOR CONTROL, ALSO PLAYS A ROLE IN COMPLEX THOUGHT PROCESSES.

Concentration Takes Work

Concentration can allow you to control some of the things your brain does. Try patting your head with your right hand and rubbing your stomach in a circular motion with your left hand at the same time. At first, you may not be able to do this. Your brain is able to tell your hands to do both kinds of motion, but usually only one at a time. You can do both patterns at the same time, but to do so you must concentrate very hard.

A Brainteaser

Why do we feel dizzy after we spin around? There is a special liquid in our inner ear that helps us control our balance. When we spin our bodies, that liquid begins to move. After we stop spinning, the liquid continues to move. The movement of this liquid fools the brain. We feel dizzy because the brain thinks the body is still spinning.

The Heart

The heart is a powerful pump that pushes blood to all parts of the body through tubes called **vessels**. If all the blood vessels in your body were laid out end to end they would extend for thousands of miles.

The heart has four chambers, two that pump the blood away from it (called the **atria**) and two that accept the blood back again (called the **ventricles**). When blood is pushed from the heart, valves in the ventricles shut, making the thumping sound of your heartbeat. Even though the body has only about five quarts (4.7 liters) of blood in it, the average human heart pumps roughly 1,250 gallons (4731.3 liters) of blood every day.

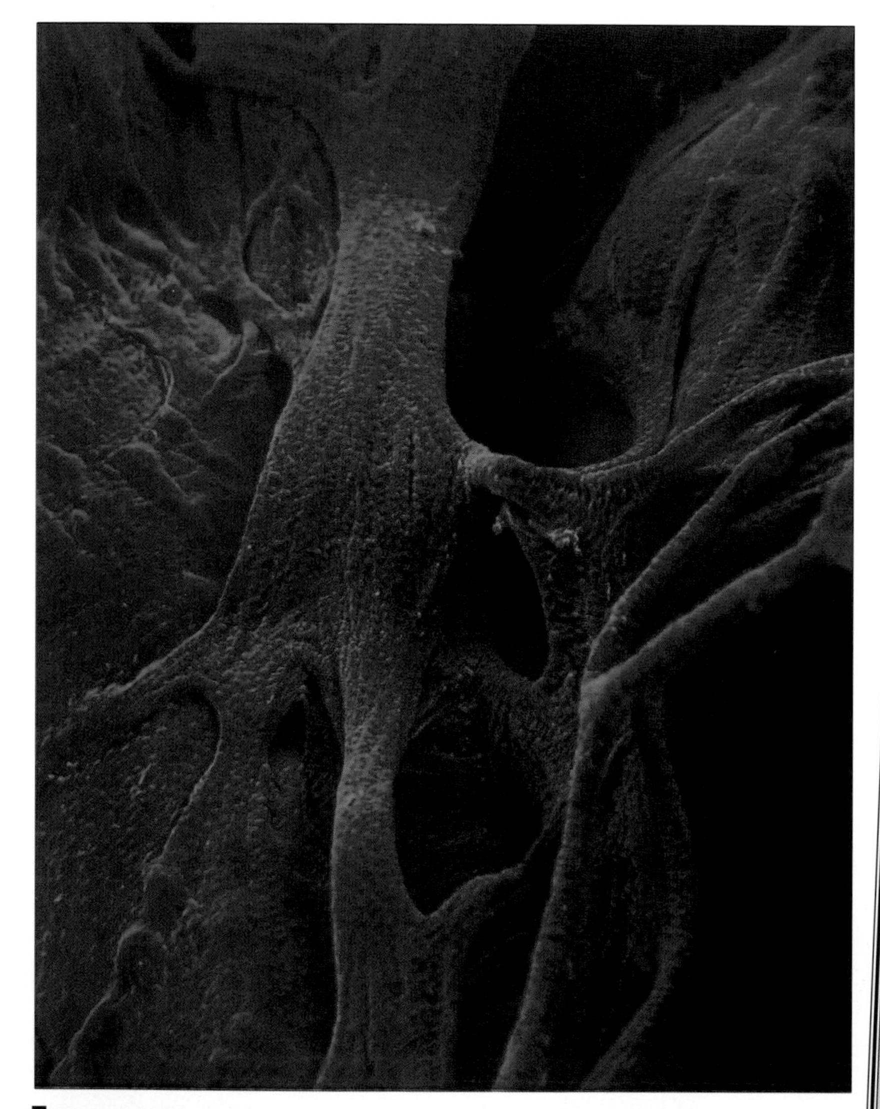

THIS ELECTRON MICROSCOPE IMAGE SHOWS A HEART PAPILLARY MUSCLE, WHICH IS RESPONSIBLE FOR OPENING AND CLOSING THE VALVES IN THE HEART.

Testing Your Pulse

Blood flows from the heart to all organs of the body through a system of thin tubes, called blood vessels. Blood vessels include **arteries, veins**, and **capillaries**. Some of these vessels are deep within the body. Others, like those in your wrist and neck, are close to the body's surface. When you measure your pulse, you measure how often your heart beats in a minute.

You will need: a bit of modeling clay and a match.

1. Place your index and third fingers on your neck just below your jaw. You can feel the blood moving.

2. Now, with your palm side up, put a tiny bit of clay on your wrist just below the thumb.

3. Stick the match into the clay so that it stands upright. The match will vibrate as the blood flows in your wrist beneath it.

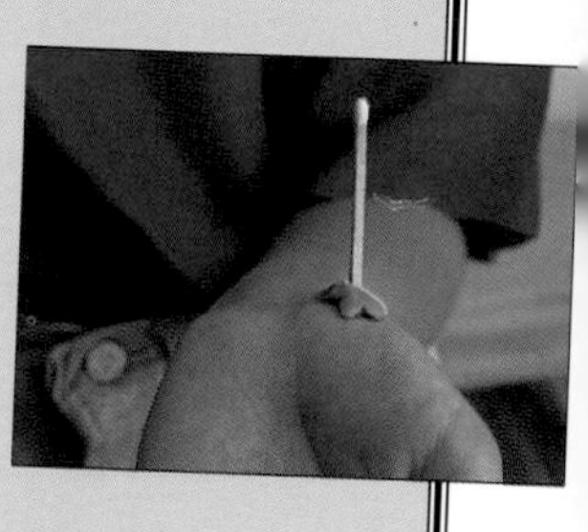

4. You can discover your pulse rate by counting how many times the match vibrates in one minute. The number you get is the amount of times your blood pumps throughout your body in just one minute. For adults, the average pulse rate is 60 to 80 beats per minute. For children, it is 80 to 140 beats per minute.

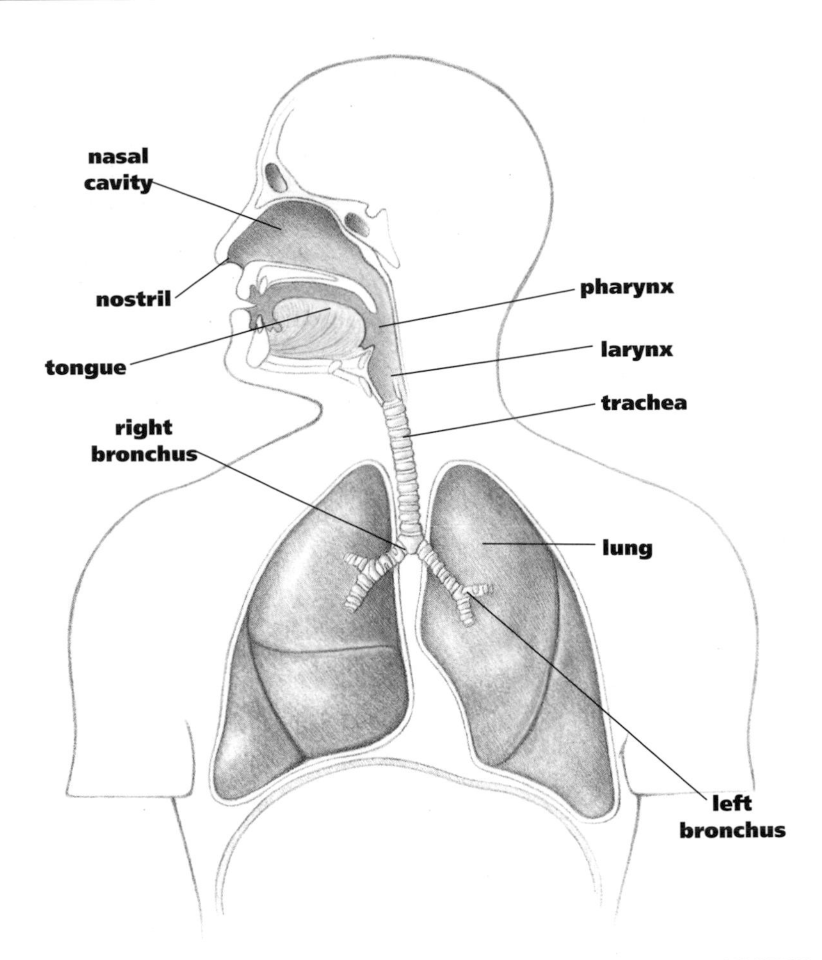

THE LUNGS ARE PART OF THE RESPIRATORY SYSTEM, WHICH ALSO INCLUDES THE NOSE, THE **PHARYNX**, THE **LARYNX**, THE TRACHEA, AND THE BRONCHI.

The Electron Microscope

Microscopes help researchers study cells and other things that are too small to see with the naked eye. One kind of microscope, called the **scanning electron microscope**, can make something appear 10 to 100,000 times larger than it actually is.

The Lungs

All the cells of our body need oxygen for nourishment. The lungs are big spongy sacs that allow us to take in oxygen so that it can be used by the cells. When we breathe, air is drawn in through the nose or mouth and down the **trachea**, or windpipe. The trachea branches off below into two smaller tubes called **bronchi**. One bronchus goes to each lung, where oxygen is absorbed by the blood and sent through the **pulmonary veins** to the heart, which sends it to all other parts of the body. The blood that comes to the lungs from the heart travels through the **pulmonary artery** and contains **carbon dioxide**, a waste product that the lungs get rid of.

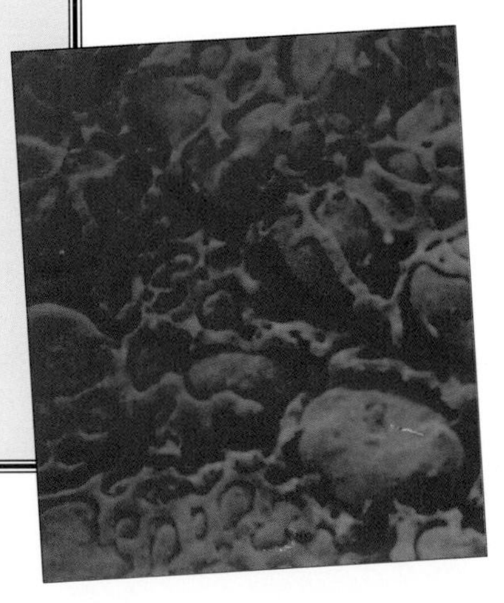

TAKEN WITH AN ELECTRON MICROSCOPE, THIS PHOTOGRAPH SHOWS CAPILLARIES WRAPPED AROUND **ALVEOLI**, TINY AIR SACS IN THE LUNGS; BLOOD FLOWING THROUGH THE CAPILLARIES RELEASES CARBON DIOXIDE INTO AND RECEIVES OXYGEN FROM THE ALVEOLI.

From Cells to Systems

Below, right: The circulatory system is the pathway of the blood; like all the body's systems, it includes different parts (cells, tissues, and organs) that work together. Below, left: The main organ of the circulatory system, the heart is also one of the most important organs in our body—it even features specialized muscle tissue. The heart is a pump that constantly circulates the body's blood supply: it sends blood containing oxygen from the lungs to the rest of the body and takes blood carrying carbon dioxide (which is the waste product left after the oxygen is removed from the blood by hungry cells) to the lungs, where it is exhaled.

The heart contains four different chambers, two of which are used to process the oxygenated blood and two of which are used to process the deoxygenated blood; this is important because if the two types of blood become mixed together, the result can be fatal. Oxygenated blood flows from the lungs through the pulmonary veins to the **left atrium**, then to the **left ventricle**, through the aorta to the arteries, and finally to the small capillaries, from which oxygen is transferred to the cells; in reverse, the deoxygenated blood travels from the cells through the capillaries to the veins and on to the vena cava, which deposits the used blood into the **right atrium**, from where it flows into the **right ventricle**, and then out through the **pulmonary artery**, which takes the blood back to the lungs, beginning the process all over again.

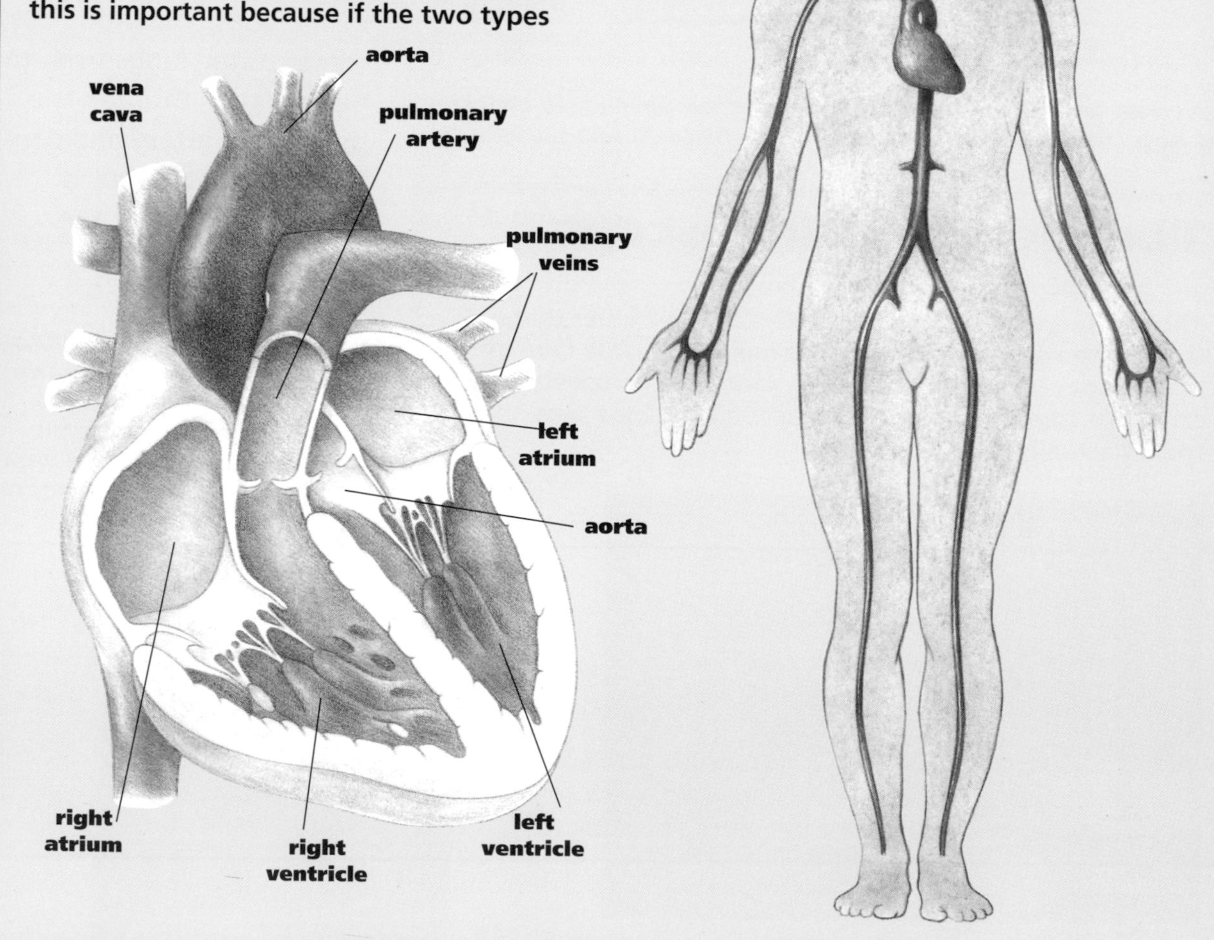

ORGANS WORKING TOGETHER

All parts of the body must work together to keep us alive and healthy. Certain groups of organs working together make up the body's **systems**. The brain is part of the **nervous system**, the heart is part of the **circulatory system**, and the lungs are part of the **respiratory system**. (Even though the heart and lungs work together to bring oxygen to the cells, they do not belong to the same system.)

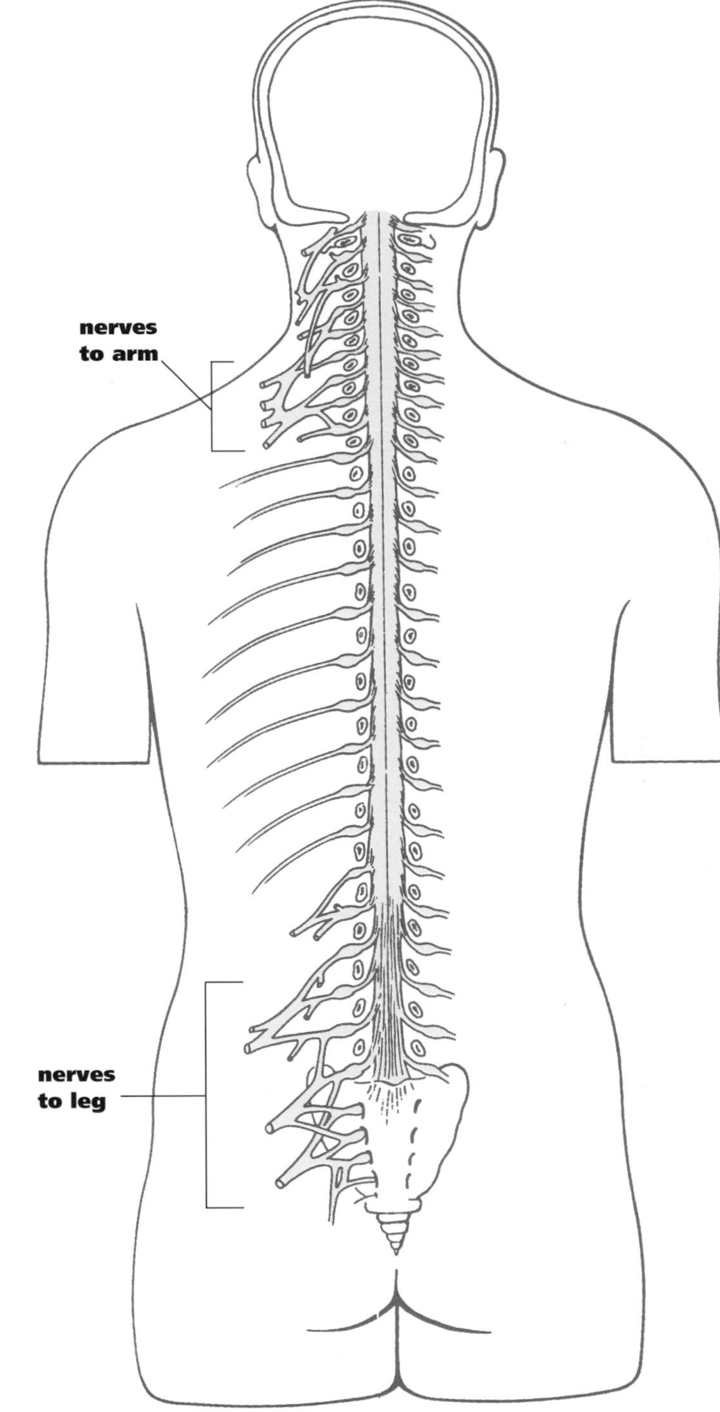

SPINAL NERVES ARE PART OF THE SPINAL CORD. SITTING ATOP THE SPINAL CORD, THE BRAIN SENDS MESSAGES TO ALL PARTS OF THE BODY ALONG THIS PATHWAY OF NERVES. THIS DIAGRAM SHOWS WHERE SOME OF THOSE NERVES LEAD.

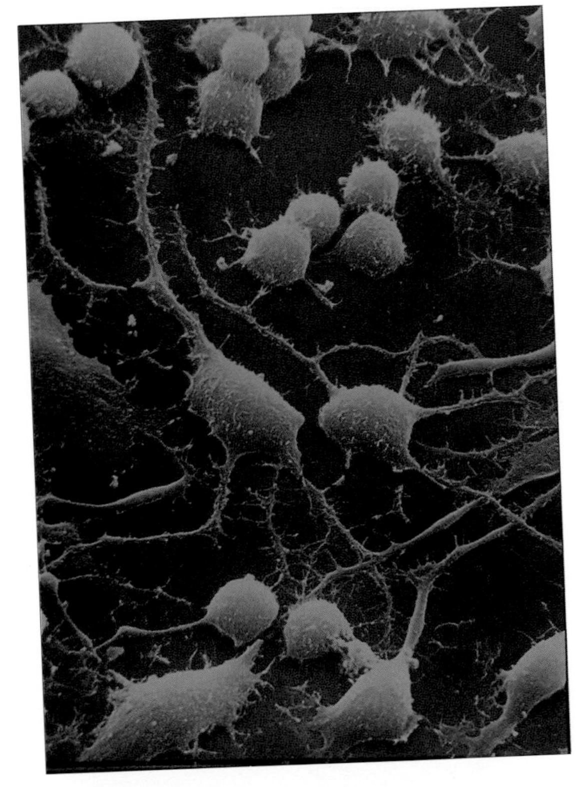

NERVE CELLS (SEEN HERE UNDER AN ELECTRON MICROSCOPE), CALLED **NEURONS**, MAKE UP ALL OF THE BODY'S NERVES AND MOST OF THE BRAIN AND SPINAL CORD.

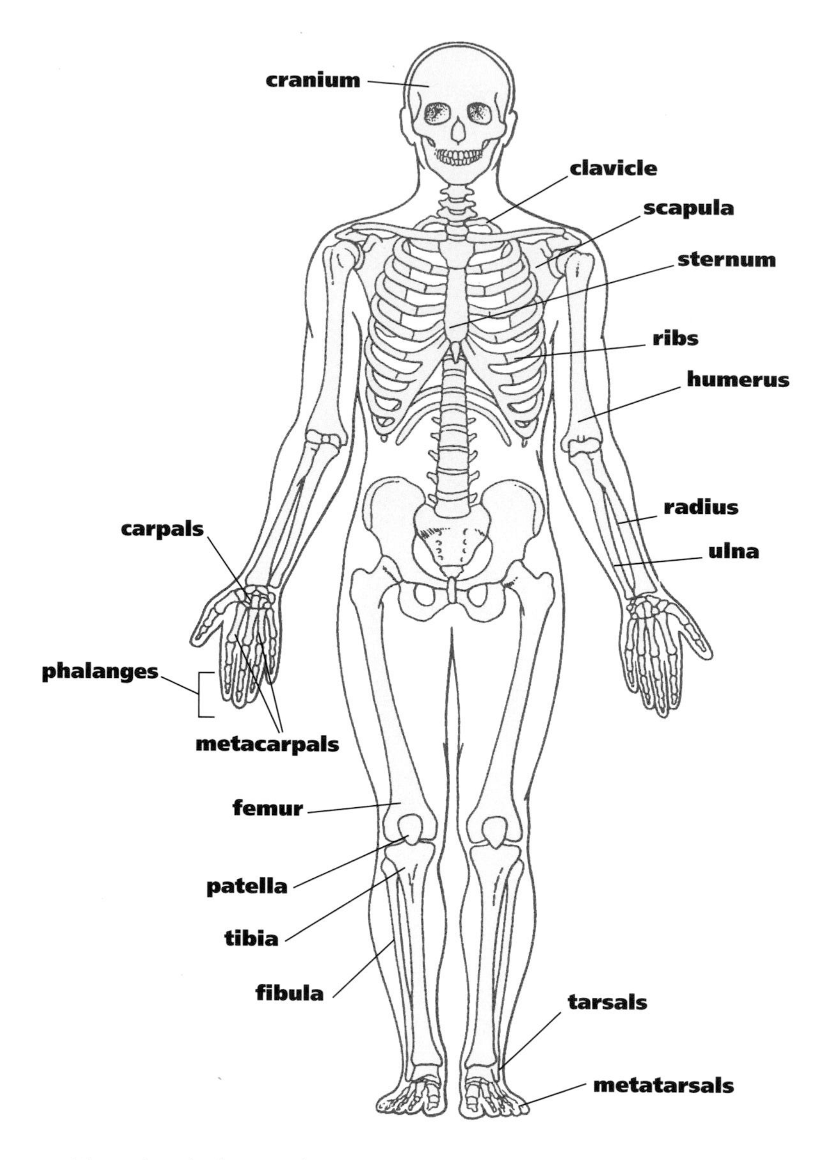

THIS ILLUSTRATION HIGHLIGHTS THE MAJOR BONES OF THE SKELETAL SYSTEM. EVERY ONE OF THE BONES IN THE HUMAN BODY HAS A NAME.

Your body has other systems, as well. Your **skeletal system** has bones that give your body a framework and protect your internal organs. Your **muscular system** has muscles that provide the movement of your bones and of certain organs. The **urinary system** cleans your blood of certain impurities and keeps your body from retaining too much water.

The **endocrine system** sends chemicals called **hormones,** which the body uses in many ways, into the blood. The **digestive system** provides energy from the foods we eat and then gets rid of wastes. Babies are born (and indeed, the species is able to endure and survive) because of the **reproductive system,** and we can see, hear, taste, feel, and smell thanks to our **sensory system** (which is closely linked to the nervous system).

How Do Broken Bones Heal?

The average adult body contains 206 bones. Bones, like all the body's organs, are made of living tissue.

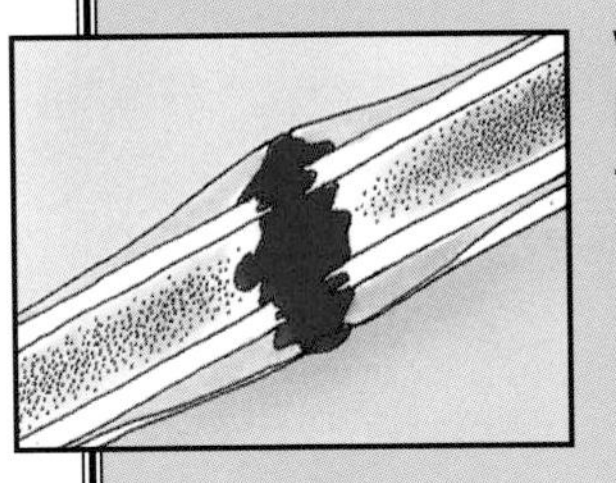

When a bone breaks, blood flows between the pieces and forms a clot.

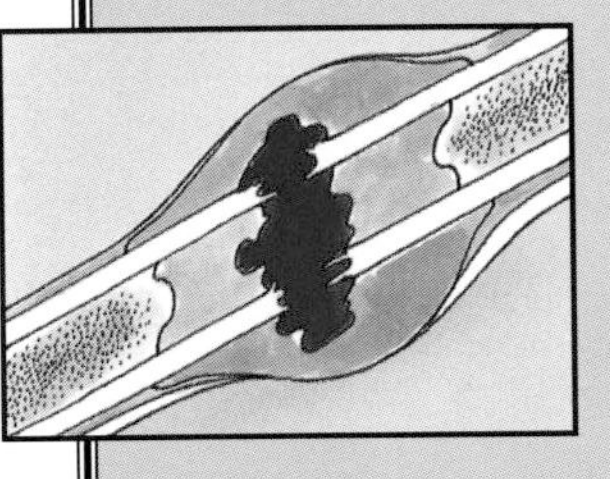

A few days later the clot is replaced by bone cells that clump around the broken area to keep the pieces together.

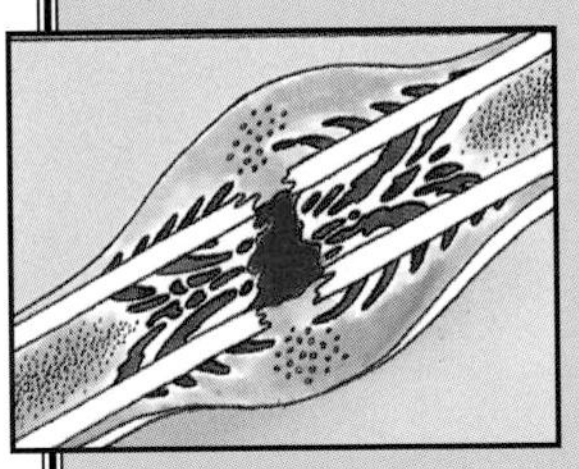

In time, the bone cells mature and the swelling goes down.

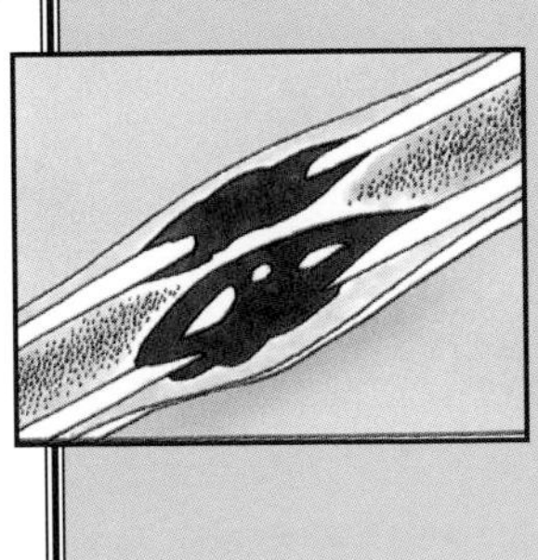

Eventually, the bone returns to its original shape. When this happens, the bone is healed.

How to Soften a Bone

Your skeletal system is the framework of your body. Your bones are sturdy and strong because they must form a scaffold for your muscles and skin, and a protective cage for many of your internal organs. Minerals keep the bones hard. This easy experiment will show you what happens to a bone when the minerals are removed.

You will need: one uncooked wing bone from a chicken, one jar with a lid, and a bottle of white vinegar.

1. Clean the meat off the bone and let it dry overnight.

2. After it is dry, put the bone in a jar filled with vinegar, put the lid on the jar, and let the bone soak for a week.

3. After a week, take the bone out of the jar and wash it off. Each day, for the next few days, try bending the bone back and forth. You will find that the bone has become soft and rubbery because the vinegar has dissolved the minerals in it.

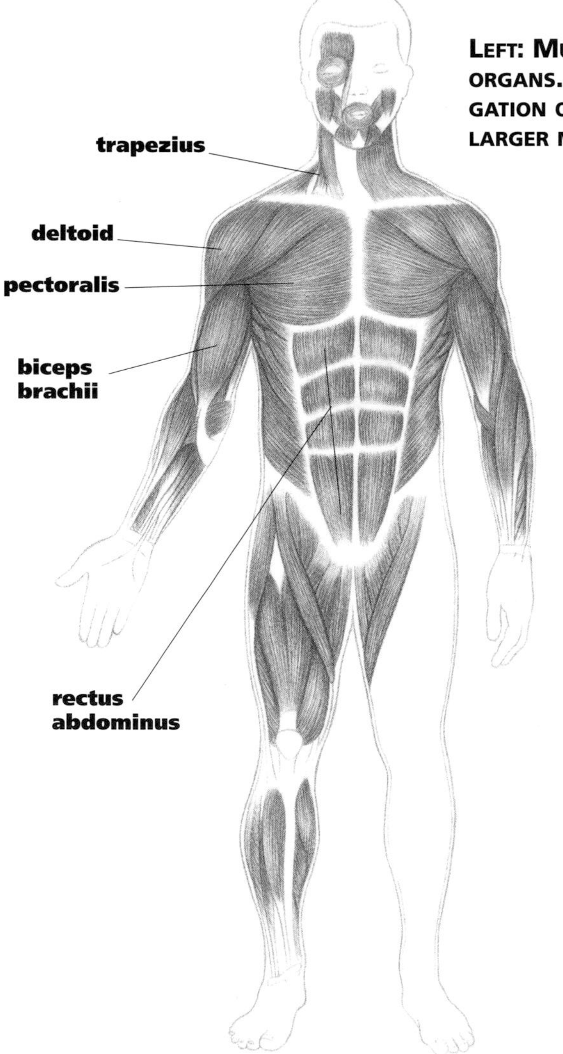

LEFT: MUSCLES ALLOW THE MOVEMENT OF THE BONES AND OF CERTAIN ORGANS. THIS MOVEMENT IS CAUSED BY THE CONTRACTION AND ELONGATION OF MUSCLE CELLS. THIS DIAGRAM HIGHLIGHTS SOME OF THE LARGER MUSCLES OF THE FRONT OF THE BODY.

BELOW: FIBERS OF SKELETAL MUSCLE ARE SEEN HERE UNDER AN ELECTRON MICROSCOPE. (SKELETAL MUSCLE IS CONNECTED TO BONE.) WHEN THESE MUSCLE FIBERS CONTRACT, THEY SHORTEN AND THICKEN, CAUSING THE BONES TO WHICH THEY ARE ATTACHED TO MOVE.

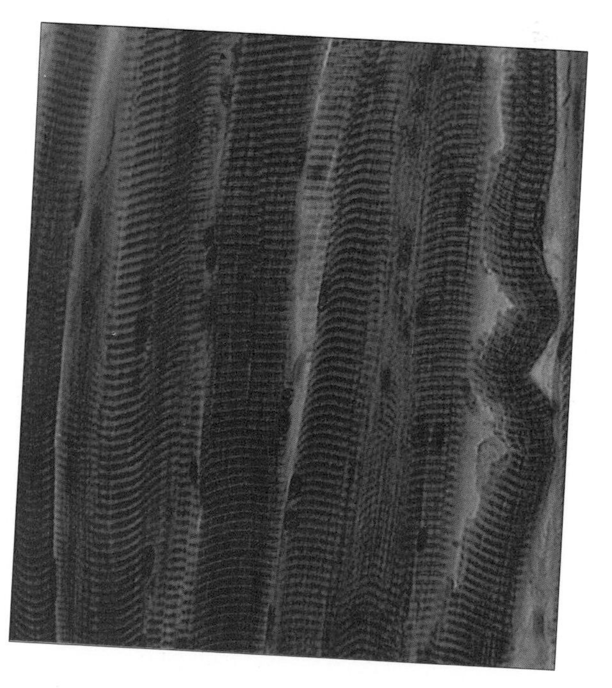

CHAPTER TWO

OUR COUSINS, THE APES

Endangered Apes

Apes are an **endangered species.** This means that if they are not protected they will soon become extinct.

PEOPLE WHO STUDY PEOPLE

The study of human beings is called **anthropology**, from the Greek word *anthropos*, meaning "man." Anthropologists try to learn more about our species' past, present, and future in several different ways. **Cultural anthropologists** explore the customs and daily lifestyles of living peoples. **Archaeologists** are anthropologists who discover things about peoples of the past by studying things those people left behind. Some anthropologists focus on the study of languages to see what can be learned from the way people communicate with one another. **Physical anthropologists** focus on learning about the physical changes that have taken place in the human body as our species evolved. Sometimes physical anthropologists study the behavior of monkeys and apes in the hopes of learning something about the way our ancestors behaved millions of years ago. At that time, our ancestors were very primitive and probably behaved more like monkeys and apes than they did like modern humans.

ARCHAEOLOGISTS AT WORK IN WYOMING.

The Four Fields of Anthropology

Cultural Anthropology: The study of the customs and lifestyles of living peoples.

Physical Anthropology: The study of human evolution and of primates.

Linguistic Anthropology: The study of human languages.

Archaeology: The study of peoples of the past.

TOP: A BLACKSMITH USES TOOLS TO CRAFT A HORSESHOE. ABOVE: A CHIMPANZEE USES A STICK TO PULL TERMITES FROM A MOUND.

OUR CLOSEST LIVING RELATIVES

Monkeys, apes, and humans are **primates**. Scientists have put all three into the same category because we have certain things in common, from the way our bodies are structured to the way we take care of our young. All primates, for example, have fingers and toes instead of paws, and nails instead of claws. All primates have a very strong bond between mothers and their infants.

Unlike monkeys, apes do not have tails and are generally more intelligent. Apes can recognize themselves in mirrors, while monkeys react to a mirror image as if they are looking at another animal. In fact, apes are our closest living relatives. They are more like humans than any other animal alive today.

There are four kinds of apes: the **gibbons**, the **orangutans**, the **gorillas**, and the **chimpanzees**. Of them all, chimps are most similar to humans in the chemical makeup of their bodies and in the way they behave. For example, chimps in the wild have been known to make tools. To get a favorite food (termites), a chimp will sometimes find a twig, strip off its leaves, and stick it into a termite mound. Then the chimp will pull the stick out and eat the termites clinging to it. Chimps have also been known to get water

A YOUNG CHIMP HUGS ITS MOTHER.

from shallow pools by making a sponge out of chewed leaves, soaking up the water with this sponge, and then sucking the liquid out of it. Chimps also use leaves to pat wounds.

Probably the most humanlike feature of chimps is the way they communicate with each other and with humans. They hug and kiss, as humans do, and sometimes get depressed if their mothers or children die.

Because they are so intelligent and have hands instead of paws, chimps and gorillas have been able to learn American Sign Language. Using sign language, they are able to communicate simple ideas to their human teachers. More recently, researchers have been using computers in their work with apes.

Why Can't Apes Learn Human Speech?

Some researchers have been able to teach apes to make certain sounds that are like human words. But only humans can make all the sounds necessary for speech. This is because an ape's neck is built differently than a human's. The larynx, or voice box, is the part inside an animal's neck that allows the animal to make sounds. In apes, as in most mammals, the voice box sits high up inside the neck. But in humans, the voice box is lower, allowing us to make more sounds and to speak as we do.

Close to Human

In the chemical makeup of their bodies and in the way they behave, chimpanzees are more like humans than they are like monkeys. Chimpanzees have about 98 percent of the same DNA that is found in humans, and they can be infected by human diseases.

BECAUSE CHIMPANZEES ARE SIMILAR TO HUMANS IN SO MANY WAYS, THERE IS MUCH THAT CAN BE LEARNED ABOUT OUR ANCIENT ANCESTORS FROM STUDYING CHIMPANZEES' BEHAVIOR AND PHYSICAL MAKEUP. THESE PHYSICAL AND MENTAL SIMILARITIES ALSO MAKE IT POSSIBLE FOR SCIENTISTS AND CHIMPANZEES TO COMMUNICATE.

Talking With Apes

Teaching **American Sign Language**, or **Ameslan** for short, to an ape takes a lot of time and patience. But researchers who do this usually get good results from their primate students. The first researchers to try this were **Beatrice and Allen Gardner** in 1966. At the end of four years, they had taught a young chimpanzee named **Washoe** to "talk" using over 130 signs. Since this first experiment, many other chimps and gorillas have been taught to use hundreds of signs to communicate with their teachers. Sometimes the animals answer simple questions, sometimes they identify objects, and sometimes they ask for things. One of the most famous apes to learn Ameslan was **Koko**, a gorilla who worked with **Dr. Francine Patterson** (both pictured above). One day, Koko asked Dr. Patterson for a kitten. When Dr. Patterson gave Koko a toy kitten, Koko was not satisfied. She was only happy when Dr. Patterson let her choose a real kitten from a litter of newborns. Koko named the kitten **All Ball** and treated it like it was her baby. She would carry Ball tucked in her thigh, dress Ball up with cloth napkins, and even sign to the kitten that they should tickle each other.

Besides American Sign Language, apes have been taught to communicate using symbols of different colored plastic pieces and simple computer programs.

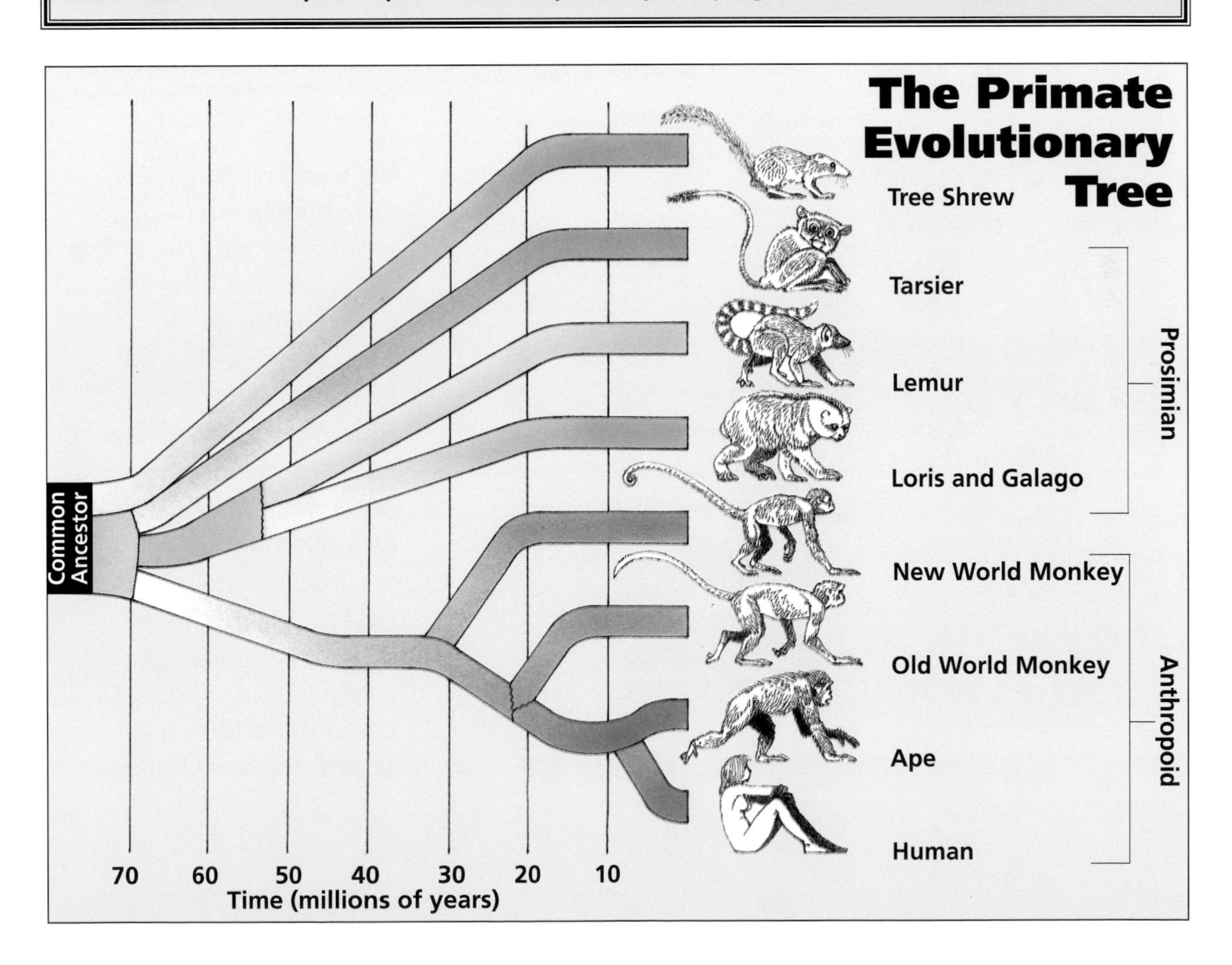

The Human's Place in Nature

Taxonomists are scientists who categorize different organisms according to the organisms' similarities and differences (the science of **taxonomy** was first devised by Carolus Linnaeus). The basic sequence of categories, from most general to most specific, is: **kingdom, phylum, class, order, family**, genus, and species. The diagram below shows how humans are classed in taxonomic terms.

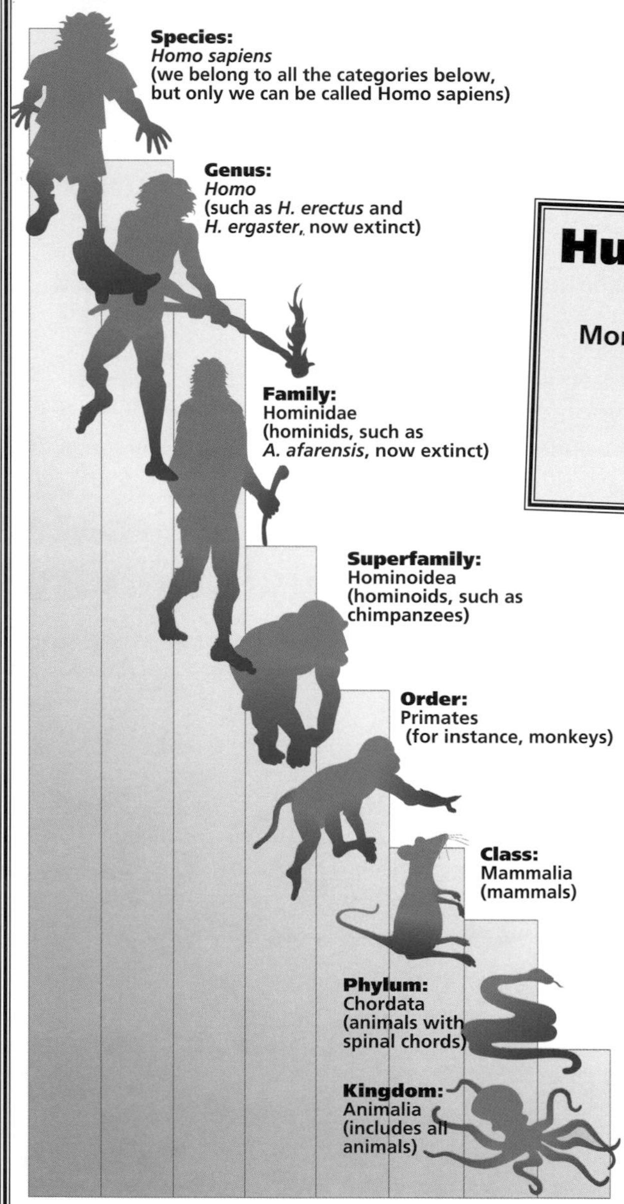

Species:
Homo sapiens
(we belong to all the categories below, but only we can be called Homo sapiens)

Genus:
Homo
(such as *H. erectus* and *H. ergaster*, now extinct)

Family:
Hominidae
(hominids, such as *A. afarensis*, now extinct)

Superfamily:
Hominoidea
(hominoids, such as chimpanzees)

Order:
Primates
(for instance, monkeys)

Class:
Mammalia
(mammals)

Phylum:
Chordata
(animals with spinal chords)

Kingdom:
Animalia
(includes all animals)

APES AND HUMANS

The theory of evolution does not tell us that we are descended from apes—rather, apes and humans share a common ancestor. This means that many millions of years ago a species of animal existed that was the ancestor of both apes and humans. This species continued to evolve until, at some point, it branched off in different directions. One branch evolved into what are today the modern apes, and the other branch evolved into us. Humans and apes together fall into a category called **hominoids**, but humans beings alone, including our extinct ancestors, are called **hominids**.

What are the features that make us human? It is true that apes are able to make tools and learn computer and sign language, but only humans

Humans Are Hominids

Monkeys, Apes, Lemurs, Tarsiers, and Humans—**Primates**

Apes and Humans—**Hominoids**

Humans—**Hominids**

Bipeds and Knuckle Walkers

The process of evolution is responsible for the differences among primates (and all species), including structural differences. Probably the result of a change in the environment of a common ancestor, structural differences in the feet, hips, legs, and knees cause humans and apes to walk differently from each other. Humans move by walking erect on two feet (bipedalism) while apes use the knuckles of their hands to travel on all four limbs (knuckle-walking).

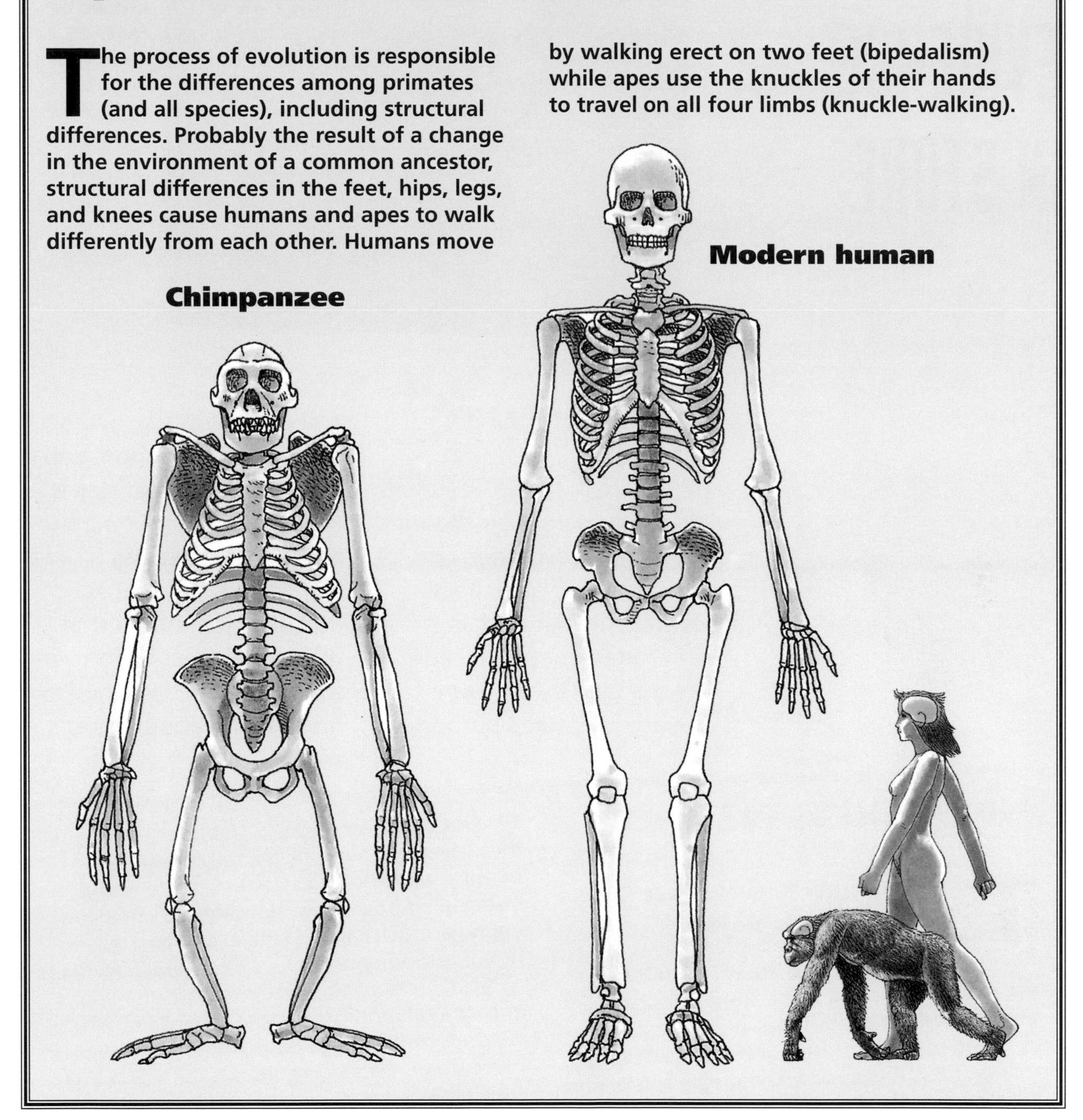

make tools from other tools, and only humans have the ability to communicate using speech.

Humans have other distinguishing characteristics. For instance, our hands are the most specialized of any animal, including such primates as chimpanzees and gorillas. Another reason humans are unique is that our brains are the most complex of any living creature. And, except for a few flightless birds, such as the penguin and ostrich, humans are the only animals that walk upright all the time. This kind of movement is called **bipedalism**.

CHAPTER THREE

EVIDENCE IN STONE

FOSSILS SUCH AS THIS ONE PROVIDE A RECORD OF PAST LIFE ON EARTH.

ARTIFACTS AND FOSSILS

Scientists base their theories of hominid evolution on the discovery of **artifacts** and **fossils**. An artifact is any object that was made or changed by someone to be used for some purpose. A spear point is an artifact, as are a hoe, a spoon, and a house. A lump of clay is not an artifact, but a bowl made from that clay is.

A fossil is physical evidence of something that was once alive. A footprint from the past hardened in mud is a fossil. An insect

How Old Is It?

There are many different ways to find out how old fossils and artifacts are. Sometimes the fossils themselves are tested, and any artifacts found with them are usually assumed to be from the same time period. At other times it is necessary to date the rock surrounding the artifacts and fossils. Then it is assumed that the fossils and artifacts are from the same time period when the rocks around them were formed.

Two popular methods of dating are **radiocarbon** (or **C14**) and **potassium/argon** (or **K/AR**) dating. Radiocarbon dating is used to determine the age of something that was once living and still has some of its original material left (for example, an artifact made from bone). This kind of dating is based on the fact that all living things have carbon throughout their bodies. When something dies, whether plant or animal, the carbon begins to decay, and scientists have found out exactly how long that takes. By measuring how much carbon is left, it is possible to figure out when the plant or animal had died.

Potassium/argon is used to date the layers of rock surrounding an artifact or fossil. Many minerals in the earth have potassium in them. Like carbon, potassium decays, and as it does it becomes a gas called argon. By measuring the amount of potassium and argon in rock, it is possible to tell when the rock was formed.

Pottery from the ancient Mimbres culture of the American Southwest. Artifacts such as these help archaeologists learn about peoples of the past.

trapped in **amber** (the hardened sap of ancient trees) is a fossil. An **imprint fossil** shows the impression of a plant or an animal's body in rock, although the plant or animal has long since decayed. Bones, teeth, shells, and wood can become fossils, too, but most of the time this doesn't happen. Usually a dead tree will simply decay, and an animal will become

A fossil plant.

How to Make a "Fossil" Cast

A **cast** is a copy of a fossil or an artifact. Sometimes museums display casts when the real fossil or artifact is unavailable or too fragile to display.

You will need: a paper cup, modeling clay, plaster of Paris, water, a spoon, and a small shell.

1. Roll modeling clay into a ball and drop the ball into the bottom of a paper cup.

2. Flatten the ball into the bottom of the cup so that it fills the bottom.

3. Press the shell into the clay firmly enough for it to leave its impression. Then remove the shell and put it aside.

4. Using the spoon, mix the plaster of Paris with water according to the directions on the package. (Remember that plaster of Paris dries very quickly. You must mix it just before you are ready to make your cast.)

5. Pour the plaster of Paris into the cup, covering the clay, until the cup is about three-quarters full. Then put the cup aside. Do not shake or move it.

6. After about an hour, peel the paper cup away from the plaster and remove the clay. You will have a cast of the shell. (If you prefer, you can use something other than a shell, such as a twig or small piece of coral. Casts of "artifacts" can be made using coins, plastic rings, or any other object small enough to fit in the cup.)

AN EARLY HUMAN DIES AND IS BURIED NATURALLY, BY WIND OR WATER PUSHING SOIL OVER THE BODY—OR PERHAPS BY THE ASHES OF AN ERUPTING VOLCANO.

a meal for scavengers. But every now and then, if the conditions are right, fossilization may happen instead.

How does a bone turn into a fossil? First, after the animal dies, it must somehow be buried. This can happen naturally in a number of ways. Winds may blow dirt over it, a flood may cover the body with mud and silt, or a volcano in the area may erupt and spread ashes over it. However it happens, the animal must be buried before other animals have completely eaten its remains and before the weather has eroded its bones.

After the animal is buried, its soft parts will decay and be eaten by bacteria and other organisms in the soil. Eventually its bones will also decay, even though bones and teeth are the hardest parts of the body. But if the bones decay slowly enough, minerals in the soil will start to fill in the bone's cells. The minerals will replace the cells as the bone decays until there is no longer any bone material left. After many thousands of years, the bone will become a fossil, which means it will turn to solid rock. This can also happen to teeth, shells, wood, and, in some rare cases, even skin. Over time, as weather erodes the ground surrounding the fossil, the buried fossil may show itself. Then it may be discovered and studied.

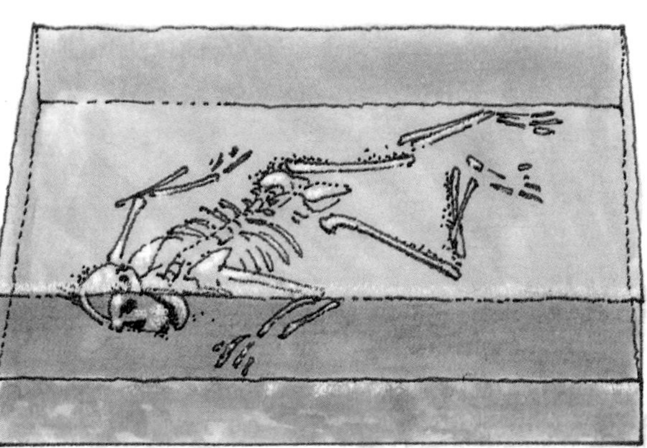

THE HUMAN'S SOFT PARTS DECAY.

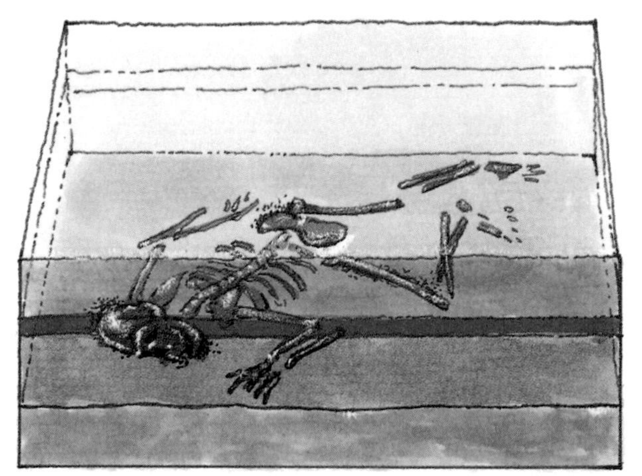

LATER, AS THE BONES SLOWLY DECAY, THE CELLS OF THE BONES ARE REPLACED BY MINERALS IN THE SOIL. IN THIS WAY, THE BONES BECOME FOSSILS.

AS WEATHER ERODES THE GROUND, THE FOSSILS MAY BE REVEALED AND DISCOVERED BY ARCHAEOLOGISTS. SUCH FINDS TEACH INVALUABLE LESSONS ABOUT THE DISTANT PAST.

RIGHT: THIS IS A FOSSIL OF A SWIMMING TURTLE THAT WAS FOUND IN GREEN RIVER, WYOMING. IT DATES FROM THE EOCENE EPOCH, WHICH IS A SUBDIVISION OF THE TERTIARY PERIOD IN THE CENOZOIC ERA. GREEN RIVER USED TO BE PART OF AN ENORMOUS FRESHWATER LAKE DURING THE EOCENE EPOCH, AND TODAY THE EXPOSED AREA OF THE ONCE-ENORMOUS LAKE IS A RICH SOURCE OF FOSSILS. BELOW: THIS IS A FOSSIL OF A FROG THAT WAS FOUND IN PROVENCE, SOUTHERN FRANCE. IT DATES FROM THE OLIGOCENE EPOCH (WHICH DIRECTLY FOLLOWED THE EOCENE EPOCH).

RIGHT: THIS IS THE FOSSILIZED SKULL AND LOWER JAW OF A SABER-TOOTHED CAT. SABER-TOOTHED CATS EVOLVED SOMETIME DURING THE OLIGOCENE EPOCH, WHICH BEGAN ABOUT 34 MILLION YEARS AGO, AND BECAME EXTINCT DURING THE PLEISTOCENE EPOCH, WHICH ENDED ABOUT 10,000 YEARS AGO. SCIENTISTS ONCE THOUGHT THAT THE SABER-TOOTHED CATS BECAME EXTINCT BECAUSE THEIR LARGE TEETH BECAME SO AWKWARD THAT THEY COULD NO LONGER KILL THEIR PREY (THIS IS THE OPPOSITE OF ADAPTATION). TODAY, HOWEVER, SCIENTISTS SPECULATE THAT THE MAIN FOOD SOURCE OF THE SABER-TOOTHED CATS DISAPPEARED, CAUSING THE MIGHTY FELINES TO DIE OUT.

SCIENTIFIC DETECTIVES

The science of finding and studying fossils is called **paleontology**. Physical anthropologists who seek our ancient hominid relatives are **paleontologists**, or **paleoanthropologists**—these scientists are searching for skeletons old enough to be fossils.

The paleontologist is a kind of detective, but instead of trying to solve a crime, he or she tries to solve some of the mysteries of the past. Paleontologists must look first for clues that tell where the evidence—fossils and artifacts—can be found. Sometimes they use electronic sensors and other equipment to help them. Sometimes they

Tools of the Trade

Paleontologists often use high-tech equipment, like computers and surveying instruments, to help them in their work. But when it comes to digging fossils and artifacts out of the ground, patience, hard work, and simple tools are best. Each object that is uncovered must be recorded, carefully removed from the earth, stored, and labeled. Some of the tools used include:

Trowels—Used to remove layers of dirt from the digging site.

Paintbrushes, Toothbrushes, Dentists' Picks—Used to remove dirt from a fossil or an artifact.

Clipboards and Paper, Cameras—Used to record the fossil or artifact by drawing it, writing a description of it, and taking a photograph of it.

Digging, or **excavating**, is only part of the paleontologist's work. After the fossils and artifacts have been discovered, they must be examined. Only when all of the evidence has been gathered together and analyzed can the paleontologist begin to form a picture of what was happening so long ago in a particular place and time.

A PALEONTOLOGIST WORKS AT AN EARLY HOMINID SITE.

accidentally find fossils poking out of hillsides and dry riverbeds that are eroding. Sometimes these detectives search places where fossils and artifacts have already been discovered in the hopes of finding more.

Hominid fossils are mostly bones and teeth. By studying them, paleontologists can discover the changes that the human skeleton has undergone over time. These scientists can also trace changes in human technology by studying the artifacts that early people made and used.

As they gather this information, paleontologists are able to piece together a picture of the past. Although this picture is hardly complete, each day new and exciting discoveries are made that add another aspect to our knowledge.

Fleshing It Out

One of the most dramatic ways to bring our past to life is by **reconstruction**. Reconstruction involves examining bones or fossil bones and figuring out what the person who owned them looked like when he or she was living. This is done by technicians using computer graphics and artists' materials.

First, the technician must make casts of the bones to be used. If he or she does not have a complete skeleton to work with, the missing pieces must be filled in when the casts are made. Then, using modeling clay, the technician adds layers of "muscle" to the bones, then "connective tissue," then "skin." Finally, the outside parts, like hair and ears, are added.

There is no way to tell from the bones what color the person's eyes, skin, and hair were, or how much hair was on the body. These details are based on educated guesses.

Talking Bones and Teeth

Human bones and teeth, whether fossilized or not, tell a lot about the people who owned them. By looking at the pelvic bone (in the hip area), it is possible to tell whether someone was male or female. (Females have wider pelvic bones because they need more room to give birth.) The age of the individual at the time of death can be discovered by looking at the skull and teeth. The ways the surfaces of the teeth are worn from chewing also give an idea of what kinds of food the person ate. If the person did certain activities a lot, like spear throwing, marks on the bones will show signs of it. It is even possible to tell whether someone was sick, because certain illnesses, like **arthritis**, also leave marks.

EXAMINING HOMINID SKULLS IN AN ANTHROPOLOGY LABORATORY.

CHAPTER FOUR

EARLY HOMINIDS

THE BEGINNINGS OF CHANGE

In the course of evolution, hominids experienced many changes in bodies and in intelligence. These changes did not happen separately; instead, as one part of us evolved, other parts were affected. For example, as the brain got bigger, our intelligence grew and we were able to figure out how to make tools. But we were able to make tools only because our hands developed, too. And our hands became more sophisticated because we had become bipedal.

Our earliest hominid relatives looked very different than we do today because they belonged to different species. In fact, paleontologists have so far identified several species of hominid that existed before us. These species are now extinct. Modern humans, called *Homo sapiens*, are the only species of human alive today.

THE EARLIEST HOMINIDS

Sometime around seven million years ago, the climate of Africa became drier. With less rain, forests began to thin out as trees died. Land that was once wooded began changing into open grasslands. At that time, there were probably primatelike animals

The Bigger the Braincase...

The braincase is the part of the skull that covers the brain. By measuring the size of the braincases of early hominids, scientists can get an idea of how large an individual's brain was. The bigger the braincase, the bigger the brain.

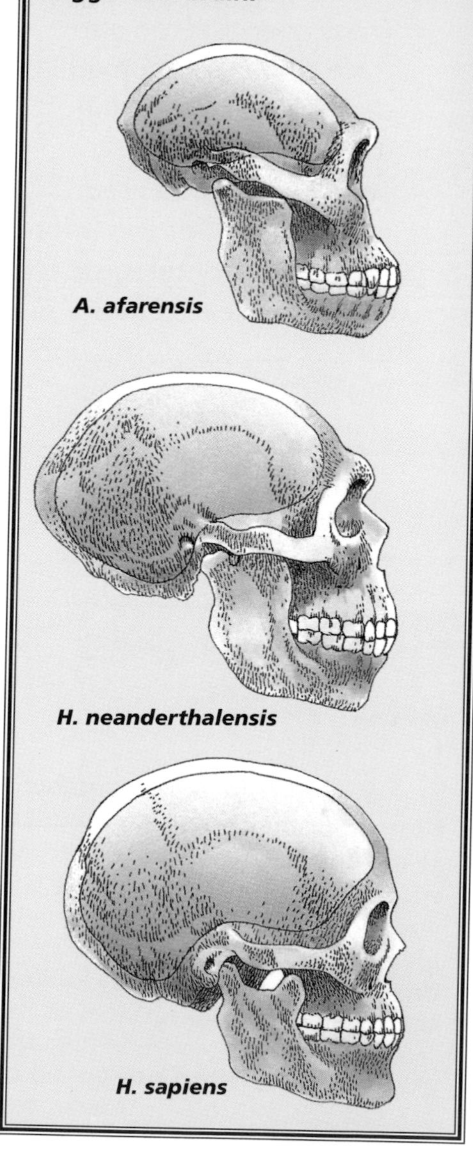

An Early Hominid Stroll

It happened about 3.5 million years ago in a place now called Laetoli in Tanzania, Africa. Two figures, maybe more, walked across an open plain in the shadow of the volcano **Sadiman**. Earlier in time, Sadiman had erupted, spewing out blasts of ash that had settled on the landscape, covering the ground in a blanket of white powder. Rain dampened the ash. The two figures walked side by side, leaving a trail of footprints behind them.

Later, the layer of ash hardened in the sun, preserving the footprints. Over time, Sadiman continued to erupt and layers of new ash covered the hardened footprints. Finally, erosion wore away the top layers of ash and the footprints began to show again. They were eventually discovered by paleontologists.

The two figures who crossed that ancient plain so long ago were members of the species *Australopithecus afarensis*, the earliest known hominid. Their footprints are a message to us from the past. They tell us that our ancestors walked the earth at least 3.5 million years ago.

living in the trees that gathered food both in the trees and on the ground. As the forests died, these animals were forced to develop a new way of life on the ground. Eventually, bipedalism became the preferred way of moving for these organisms. Over the millions of years it took for this to happen, changes in the structure of their feet, hips, legs, and knees allowed these hominid ancestors to move around more and more efficiently on two feet.

The fossils of our earliest hominid relatives date to about 3.5 million years ago. Known as *Australopithecus afarensis* (*A. afarensis* for short), this hominid lived in eastern Africa. *A. afarensis*

Name That Hominid

The name *Paranthropus* means "near-man." *Australopithecus afarensis* means "southern ape of the Afar." (The **Afar** is the place in Ethiopia where the first fossils of *A. afarensis* were found.) The name *Australopithecus africanus* means "southern ape of Africa." Neither *A. afarensis* nor *A. africanus* were really apes.

OLDUVAI GORGE, TANZANIA, IS A SITE RICH IN EARLY HOMINID FOSSILS.

was small. Adult females were about three feet six inches (1.1 meters) tall and weighed about 60 pounds (27.2 kilograms). Adult males were about four feet six inches (1.4 meters) tall and weighed about 100 pounds (45.3 kilograms). Although their arms were about as long as ours, their legs were shorter. Their hands and feet were longer than ours and also somewhat curved, an indication that they probably still climbed trees at times for shelter and to gather food. The braincase of *A. afarensis* was small, but the face was large, featuring a large, jutting jaw and big teeth.

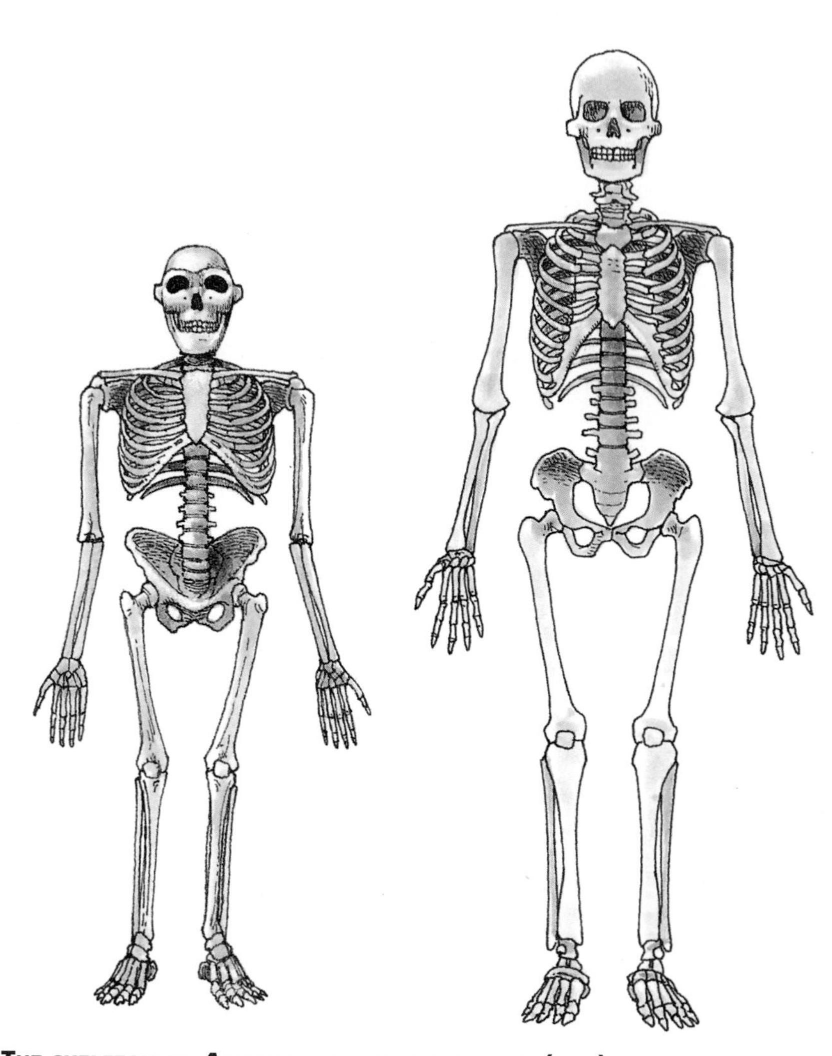

THE SKELETON OF *AUSTRALOPITHECUS AFARENSIS* (LEFT) COMPARED TO THE SKELETON OF A MODERN HUMAN (RIGHT).

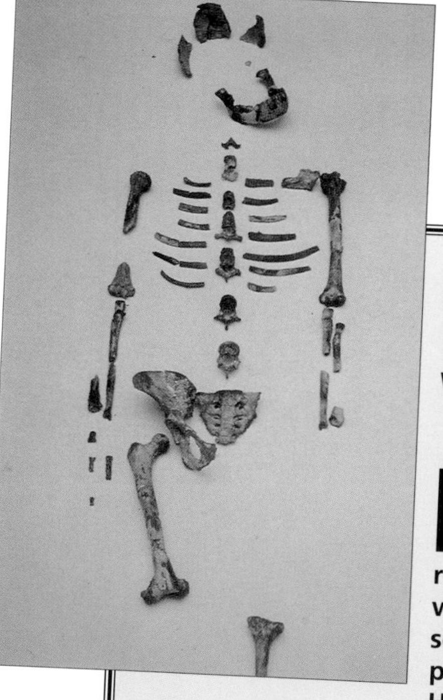

Lucy, An Important Woman

In 1976, anthropologist **Donald C. Johanson** and his research team were working under a scorching sun in the parched desert of Hadar, Ethiopia. Although they had worked all day searching for hominid fossils, all they had found were animal fossils. Then, just as the team was about to give up for the day, Johanson noticed a piece of bone eroding out of the slope above them. The bone turned out to be part of a hominid arm. As they continued to search, more fossil bones turned up. They searched for three more weeks and found hundreds of fossil bones—all of which turned out to belong to the same individual.

What they found after they had pieced all the fossil bones together was a woman three feet eight inches (1.1 meters) tall who probably weighed about 65 pounds (29.5 kilograms) when she was alive. Johanson nicknamed her **Lucy**. Lucy was an *Australopithecus afarensis* who lived about 3.2 million years ago. At the time she was discovered, hers was the most complete early hominid skeleton ever found.

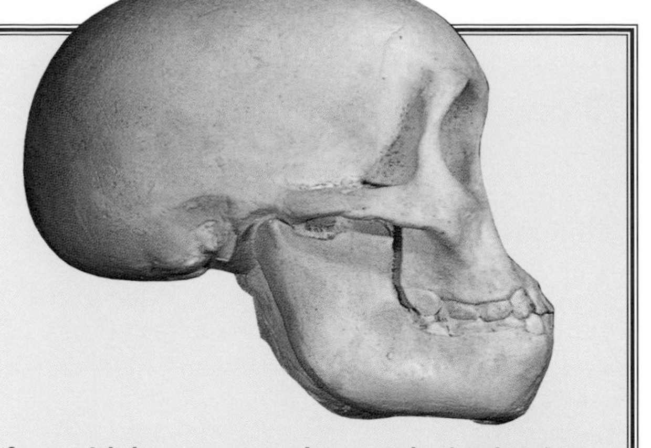

Dart's Discovery: The Taung Child

In South Africa in 1924, **Raymond Dart**, an anatomist, was getting ready to go to a wedding when one of the most important events in his career happened. Nearby, miners blasting for limestone had found fossils in a place called **Taung**. On the morning of the wedding, boxes of these fossils were delivered to Raymond Dart's home.

After the wedding reception, Dart began the work of examining the fossils. One fossil in particular caught his interest: the fossil of a small brain, called an **endocast**. The fossil brain fit into another block of limestone that Dart found in the box. He could not yet see the face that belonged with the brain because the face was hidden inside the block of limestone.

Carefully, using his wife's knitting needles, Dart chipped away at the rock until a small face with large eye sockets and a jutting jaw appeared. A small fossilized piece of the brain stem (the part of the brain that attaches to the spinal column) showed conclusively that the face belonged to a creature who walked upright.

The **Taung child** was a young hominid who lived about 2.2 million years ago and who died at the age of five or six. Dart gave the Taung child the name *Australopithecus africanus*. It was the first of this species ever to be discovered.

The Fake That Fooled the Experts

In 1912, near Piltdown, England, **Charles Dawson** found some fragments of what looked like a fossil skull. When he put the pieces together, the skull looked like a combination of an ape and a modern human. This creature was given the name **Piltdown Man**, and many people thought it was the "missing link" between our prehuman and human ancestors.

It was not until the 1950s, when new ways of dating fossils were developed, that this "fossil" was proven to be a fake. It had the skull of a modern human and the jaw of an orangutan whose teeth had been filed down. Piltdown Man was actually only about five hundred years old. Someone had stained the bones brown to make them look like fossil bones and then buried them in the ground. Today, no one knows who it was that managed to fool the scientific world for so long.

In southern Africa, between 3 and 2 million years ago, lived another hominid ancestor, *Australopithecus africanus* (*A. africanus*). *A. africanus* looked a lot like *A. afarensis*, only *A. africanus* had a slightly larger brain.

Both *A. afarensis* and *A. africanus* are called "gracile" australopithecines because of their small size. Two other species of early hominid were more heavily built than the gracile australopithecines. These hominids had bigger jaws, flatter faces, and small front teeth and large molars. Some of these

The "First Family"

A group of *Australopithecus afarensis* discovered in Hadar, Ethiopia, in 1975 have become known as the "first family." This group of at least thirteen people included adults and children. They were probably all killed at the same time by some natural disaster, such as a flash flood.

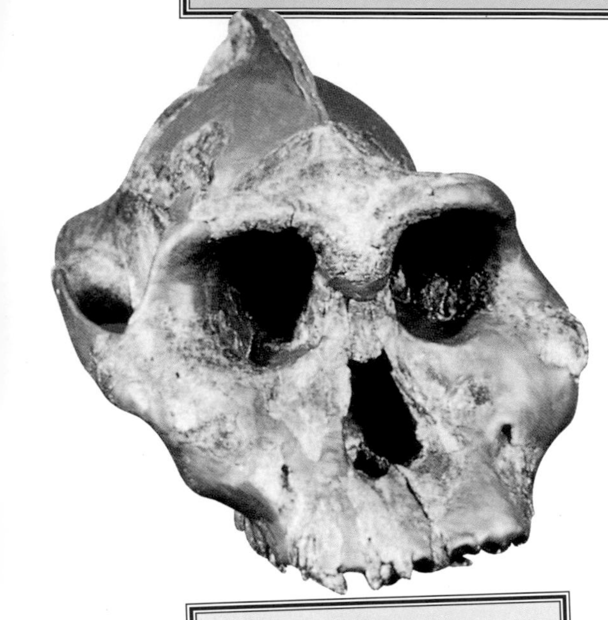

The Black Skull

Paranthropus aethiopicus is also known as the "Black Skull." It got this name because as it fossilized, much of the bone was replaced by the mineral **manganese**. Manganese is usually black, and so it gave the fossil skull a black color.

larger australopithecines had large, bony crests on the tops of their skulls, where their chewing muscles were attached. Because their jaws were so large, their chewing muscles were very thick; the crest anchored the muscles. These hominids lived in southern and eastern Africa around 2 to 1 million years ago. Today, some paleontologists call the two species *Australopithecus robustus* and *Australopithecus bosei*. Other paleontologists, however, feel that these two are from a different genus, and call them *Paranthropus robustus* (*P. robustus*) and *Paranthropus bosei* (*P. bosei*). The debate goes on.

Yet another kind of *Paranthropus*, *P. aethiopicus*, lived in eastern Africa about 2.6 million years ago. *P. aethiopicus* is unusual because its skull looks a lot like *Paranthropus*, but its face is more pushed out, like *A. afarensis*. It is believed to be the ancestor of *P. robustus* and *P. bosei*.

Although *A. afarensis* and *A. africanus* did not make tools, it is possible that *Paranthropus* did make tools out of bone and stone. This is still being debated by scientists.

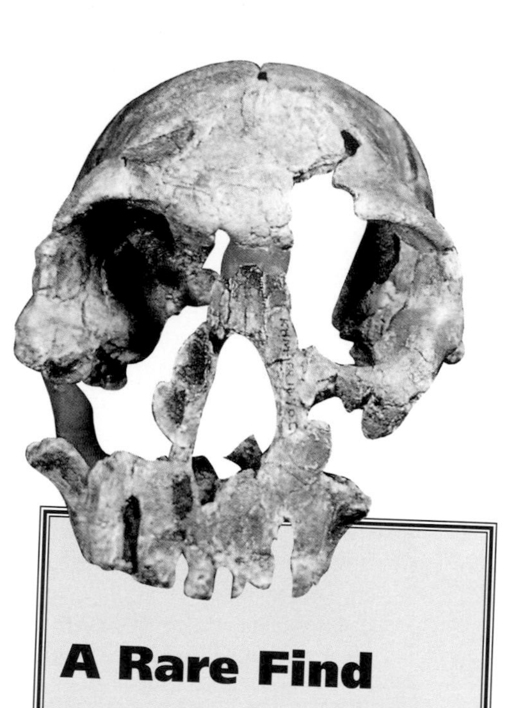

A Rare Find

One of the most famous *Homo habilis* finds is a fossil skull from eastern **Lake Turkana** in Kenya, Africa, called **KNM-ER 1470**. Most of the time, paleontologists find only fragments of early hominid fossil bones and teeth. Discovered in 1972, 1470 is a rare find because it is a nearly complete skull.

Our Sensitive Fingers

Complex brains and specialized hands allowed our early hominid ancestors to fashion tools at least 1.5 million years ago. All primates have hands, but the human hand is the most specialized. Our fingers and thumbs have many nerve endings in them and are very sensitive. Try the following experiment to find out just how sensitive our fingers and thumbs are.

You will need: two sharpened pencils, some tape, and a friend.

1. Tape the two pencils together with both points facing the same direction.

2. Have your friend close his or her eyes. Gently press both pencil points at the same time on your friend's forearm. Ask your friend how many points he or she feels.

3. Next, press the points on your friend's fingertip or thumb tip and ask the same question again. You will find that your friend only felt one pencil point on the forearm, but both points on the finger or thumb tip.

THE FIRST REAL TOOL-MAKER

Between about 2 and 1.5 million years ago, in southern and eastern Africa, lived *Homo habilis* (*H. habilis*). *H. habilis*' bones were not much different from those of *A. africanus*, except that *H. habilis* had a bigger braincase. *H. habilis* also did something very unique and important. Paleontologists know for sure that *H. habilis* made stone tools, called **Olduwan pebble choppers** by scientists.

The pebble chopper was probably an all-purpose tool used to cut and scrape. These tools were made by using one pebble to strike pieces off a larger one. This is called **flaking**. The larger one, about four or five inches (10.2 or 12.7 centimeters) long, was the pebble chopper and had sharp edges where it had been struck. *H. habilis* may have also used the flakes (the sharp smaller pieces).

These tools mark an important step forward in hominid evolution. It is the earliest evidence we have that our ancestors were beginning to learn how to make things that would help them in their daily lives.

AN OLDUWAN PEBBLE CHOPPER.

Name That Hominid

The name *Homo habilis* means "handy man." *H. habilis* got this name because he made tools.

A Fossil Treasure Trove

Olduvai Gorge in Tanzania, Africa, is a place rich in early hominid fossils. It was here that the famous husband and wife team **Louis and Mary Leakey** spent nearly twenty years searching for clues to our human ancestry. They uncovered many important finds, including the first *Homo habilis* fos-

CHAPTER FIVE

FROM THE AGE OF FIRE TO THE ICE AGE

THE FOSSILIZED SKULL OF JAVA MAN (*HOMO ERECTUS*).

THE FIRE-USERS

Homo erectus (*H. erectus*) and his close relative *Homo ergaster* (*H. ergaster*) thrived from about 2 million years ago to about 250,000 years ago. *H. ergaster* fossils have been found in eastern Africa. *H. erectus* fossils have been found in Africa and Asia. These two species were taller than earlier hominid species. Their bodies were very much like those of modern humans, but their skulls were still primitive. Although their brains were larger than *H. habilis*' brain, they were still much smaller than the brains of modern humans. *H. erectus* and *H. ergaster* also had thick,

Name That Hominid

The name *Homo erectus* means "upright man." The name *Homo ergaster* means "workman."

A First From Indonesia

In 1887, a Dutch anatomist named **Eugene Dubois** sailed to Indonesia hoping to find fossils of early humans. In 1891, on the island of **Java**, he found a braincase, a single thigh bone, some thigh bone fragments, and a tooth. Dubois named his find *Pithecanthropus erectus*, which means "upright ape-man." Today we know this species as *Homo erectus*, and this particular find as **Java Man**.

The fossils of Java Man were the first *Homo erectus* fossils to be discovered. But in Dubois' time, paleontology was not yet a respected science. Hardly anyone believed he had discovered an early form of human. People questioned his find, saying the thigh bones looked a lot like that of a modern human. Dubois was so disheartened that he buried the fossils under the floorboards of his house. He didn't publish his find until 1922—thirty-three years after his amazing discovery.

bony ridges over their eyes. *H. erectus* made stone chopping tools that are described with the adjective **Acheulean**. These tools include hand axes and cleavers. Like the pebble choppers of *H. habilis*, Acheulean tools were made by flaking—but Acheulean tools were bigger and better made. Existing Acheulean tools range from six inches (15.2 centimeters) to almost a foot (30.4 centimeters) long.

Some sites where *H. erectus* fossils were found show signs that *H. erectus* used fire. (Burned things leave behind carbon, and layers of carbon have been found at these sites.) Once these hominids had figured out how to use fire, they could then cook their food. Cooked food is softer and easier to chew than raw food. So, over many generations, *H. erectus*' teeth got smaller because they had to do less work to break up their food. As their teeth got smaller, so did their jaws. *H. erectus*' face, then, was not quite as pushed out as those of the earlier hominids.

AN ACHEULEAN HAND AXE.

STONE FLAKES.

Early Humans in China

Zhoukoudian is a place not far from Beijing, China. Here, beginning in 1920, fossils of an early human, called **Peking Man**, began to turn up. Paleontologists also found stone tools and evidence that fire was used. Peking Man belonged to the species *Homo erectus*.

The fossil remains of five individuals were uncovered in Zhoukoudian. All of the fossilized remains were broken in pieces. Anatomist **Franz Weidenreich** took this as evidence that Peking Man ate each other and that the bones had been broken during a feast. Later scientists disagreed, saying the bones were probably broken by scavenging animals or by natural forces.

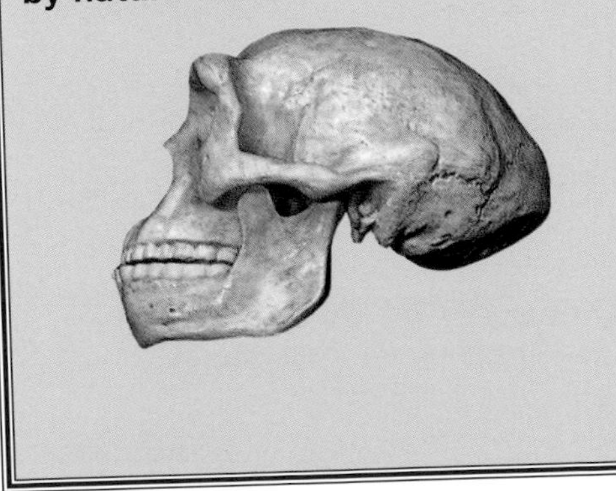

Name That Hominid

Neanderthal was named after the **Neander Valley**, Germany, where the first Neanderthal fossil was found in 1857.

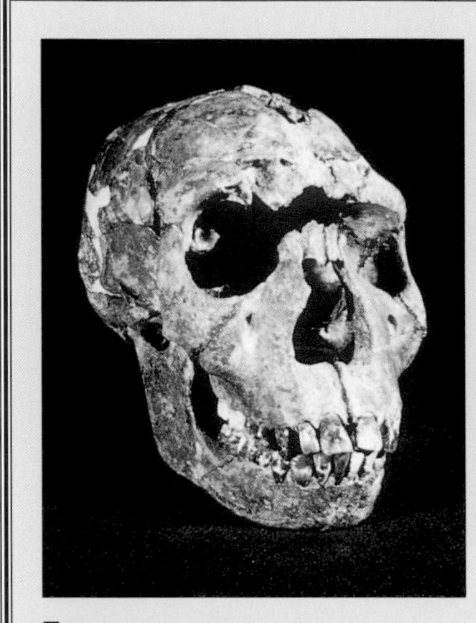

THE SKULL AND SKELETON OF TURKANA BOY.

The Boy From Lake Turkana

About 1.6 million years ago, in what is today Kenya, Africa, lived a boy who belonged to the species *Homo ergaster*. This boy died when he was about nine years old. Most of his skeleton became fossilized. When the skeleton was discovered in 1984, paleontologists gave him the name **Turkana Boy** because he was found near **Lake Turkana**. Turkana Boy was built very much like a modern human boy of about eleven to thirteen years old. He was about five feet three inches (1.6 meters) tall and probably weighed about 106 pounds (48.1 kilograms). Had he lived to adulthood, Turkana Boy might have grown to be over six feet (1.8 meters) tall.

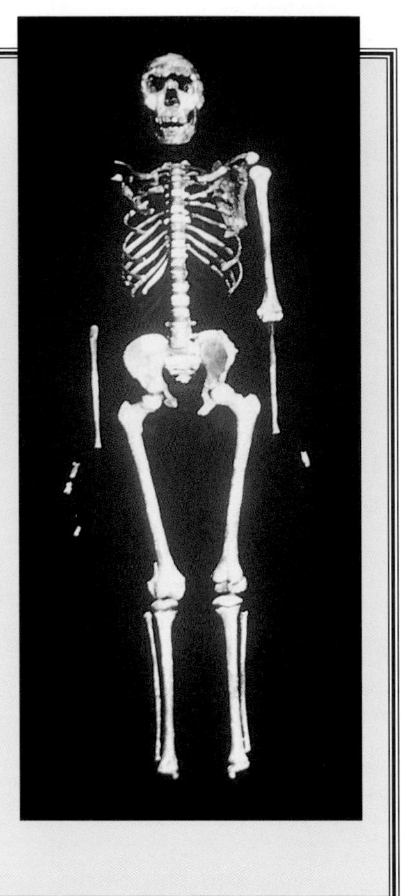

The Continents During the Ice Ages

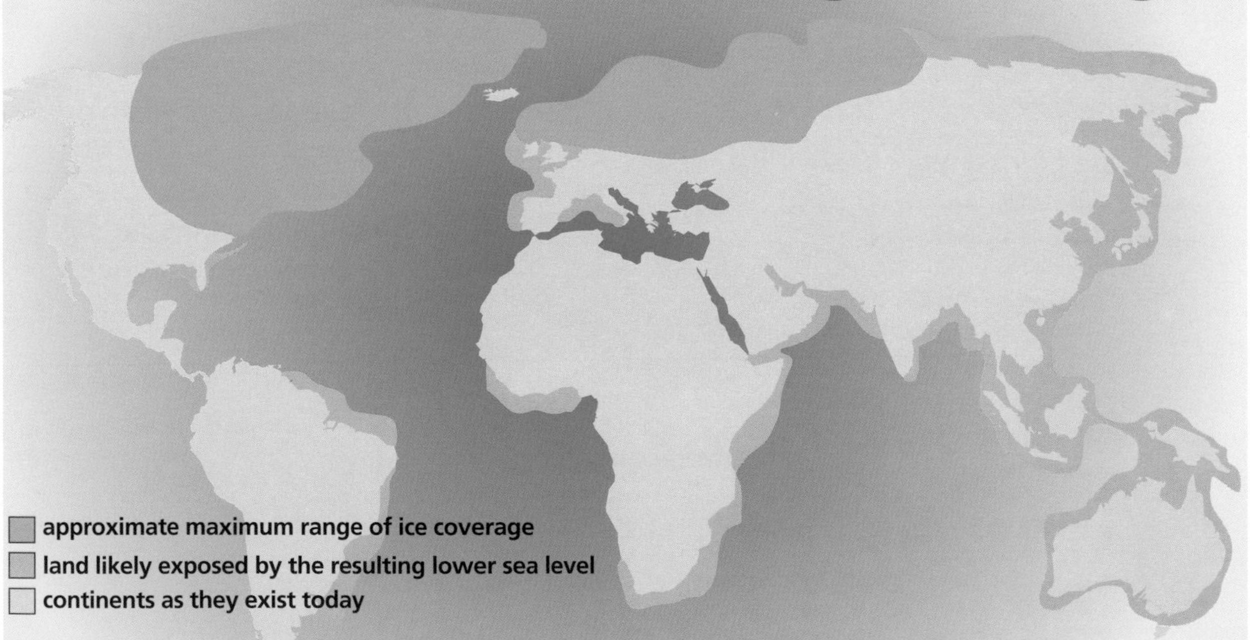

DURING THE ICE AGES, AS GLACIERS EXPANDED AND COVERED THE LAND, THE LEVELS OF THE WORLD'S OCEANS DROPPED. LAND PREVIOUSLY COVERED IN WATER WAS EXPOSED, CHANGING THE SHAPES OF THE CONTINENTS DURING EACH ICE AGE.

LIVING IN THE ICE AGES

From about 1.6 million years to about 10,000 years ago the Earth went through a series of **Ice Ages**. During this time, the Earth's climate swung between periods of extreme cold and periods of mild temperatures. It was especially cold in the northern hemisphere, where enormous glaciers built up during the cooler periods.

During this time, between about 150,000 and 35,000 years ago, the Neanderthals lived in Europe and western Asia. Neanderthals were muscular and heavily built, with thick brow ridges and a wide nose. Their brains were at least as large as ours, but were shaped differently. Their braincases were low and long. Some paleontologists think Neanderthals belonged to our own species, *Homo sapiens*, but many scientists today believe that Neanderthals belonged to a different species, *Homo neanderthalensis*.

Hominid Reconstruction

By comparing how our ancestors may have looked with how we look, scientists can learn a lot about how we evolved. (Note, for instance, how the skull enlarged over time to accommodate the growing hominid brain.) In order to get a picture of our ancestors, scientists reconstruct the faces of these early hominids based on the shape of the skulls, what is known of the hominid's environment, and so on.

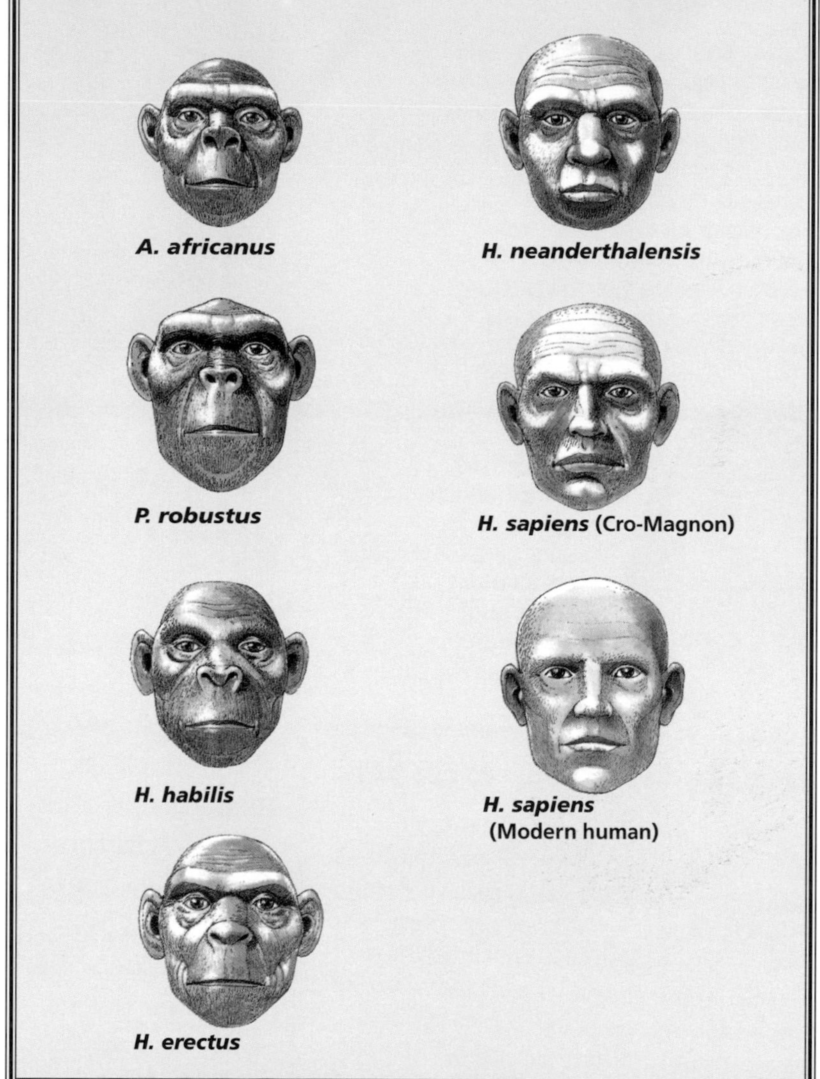

When Did Humans Begin to Speak?

All early hominid species probably had some way of communicating using sounds, but no one really knows when they began using speech. Researchers working on fossil skulls have tried to figure out how early hominid necks were most likely built. Based on this research, it is believed that the first hominid able to use simple speech was probably *Homo erectus*. But speech as we know it probably did not develop until around the time of Neanderthals.

RIGHT: THIS ILLUSTRATION SHOWS WHAT A NEANDERTHAL MAN MIGHT HAVE LOOKED LIKE. IT IS BELIEVED THAT NEANDERTHALS WERE HUNTERS WHO WERE CAPABLE OF USING DIFFERENT TOOLS AND WORE ANIMAL SKINS FOR PROTECTION. ABOVE AND BELOW: AMONG THE TOOLS THAT WERE MOST IMPORTANT TO THE NEANDERTHALS WERE HAND AXES AND SCRAPERS (MADE BY STRIKING FLAKES FROM ONE STONE WITH ANOTHER), MANY OF WHICH WERE USED TO MAKE OTHER TOOLS FROM SOFTER MATERIALS SUCH AS WOOD.

A Neanderthal Burial

In 1908, an important discovery was made at **La Chapelle-aux-Saints**, France. Three French priests found an almost complete skeleton in a grave made by Neanderthals. The man had been buried in a shallow grave with a bison leg on his chest. His arms and legs had been bent, and his head had been placed on a pillow of stones. In the grave were stone tools and broken animal bones.

Finds like this tell paleontologists that Neanderthals were the first humans known to bury their dead. The find offered clues that, for perhaps the first time in our ancestry, humans were beginning to think about the concept of an afterlife.

Neanderthals were skilled toolmakers. Like other hominids before them, they made stone tools by flaking. The scrapers, spear points, and hand axes they made are described with the adjective **Mousterian**. By carefully looking at the wear on Mousterian tools, paleontologists can tell that Neanderthals used them to make other tools out of wood. In fact, at a Neanderthal site in Germany, a sharpened spear point was found in the fossilized ribs of an elephant. Wear on some of the tools also show that they were used to scrape hides. This tells us that Neanderthals probably wore animal skins as clothing.

Nurturing Neanderthals

In 1951, **Ralph Solecki** began excavating near the town of **Shanidar**, Iraq, and soon found some remarkable things. Nine Neanderthals—seven adults and two babies—had made their home in a cave. Four of the individuals had been buried. On one grave there was evidence of pollen, which means that flowers of some sort had been placed there.

But Neanderthals not only cared for their dead, they cared for the living as well. Many of the skeletons at Shanidar show signs that these individuals survived serious injuries. One of the skeletons belonged to a man about forty years old—an old man for that time. Evidence shows that while alive he had suffered from a crippled arm, arthritis, and injuries to the head and face and had probably been blind in one eye. This evidence tells paleontologists a lot about how Neanderthals lived. It shows that taking care of the sick was important to them. It shows that Neanderthals cooperated with one another to take care of family and friends.

THESE NEANDERTHAL SKULL AND BONE FRAGMENTS WERE PHOTOGRAPHED BEFORE BEING REMOVED FROM THE GROUND.

CHAPTER SIX

THE STORY CONTINUES

Rock of Ages

The word **Paleolithic** means "old stone age" (**paleo** means "old" and **lithic** means "stone"). **Neolithic** means "new stone age" (**neo** means "new"). These words describe periods of time before people used metal to make tools.

Name That Hominid

Cro-Magnon was named after the Cro-Magnon rock shelter in France. Early human skeletons from about 30,000 to 25,000 years ago, along with extinct animal fossil bones, were found there in 1868. Cro-Magnon is the name for early *Homo sapiens* of Europe. The name *Homo sapiens* means "wise man."

MODERN PEOPLE

When did human beings become fully modern *Homo sapiens* with skeletons and skulls like ours? This is a hard question to answer because the fossil evidence from that time period is incomplete. In Asia and Australia modern humans may have existed around 40,000 years ago. But there is evidence from Africa that tells us *H. sapiens* may have existed on that continent as early as 120,000 years ago.

Eventually, around 40,000 years ago, *H. sapiens* entered Europe from wherever they originated. It is in Europe that we find the most complete fossil evidence of early modern man. Early *H. sapiens* of Europe are called **Cro-Magnon**. The fossils and tools from western and central Europe date from about 35,000 to 10,000 years ago, a period of time called the **Upper Paleolithic**. Cro-Magnon shared the continent with Neanderthals for a few thousand years before the Upper Paleolithic began. By 35,000 years ago the Neanderthals were gone. We are not sure what happened to them. Maybe they had to compete with Cro-Magnon for food and other resources and did not succeed. Maybe when the Cro-Magnons entered Europe

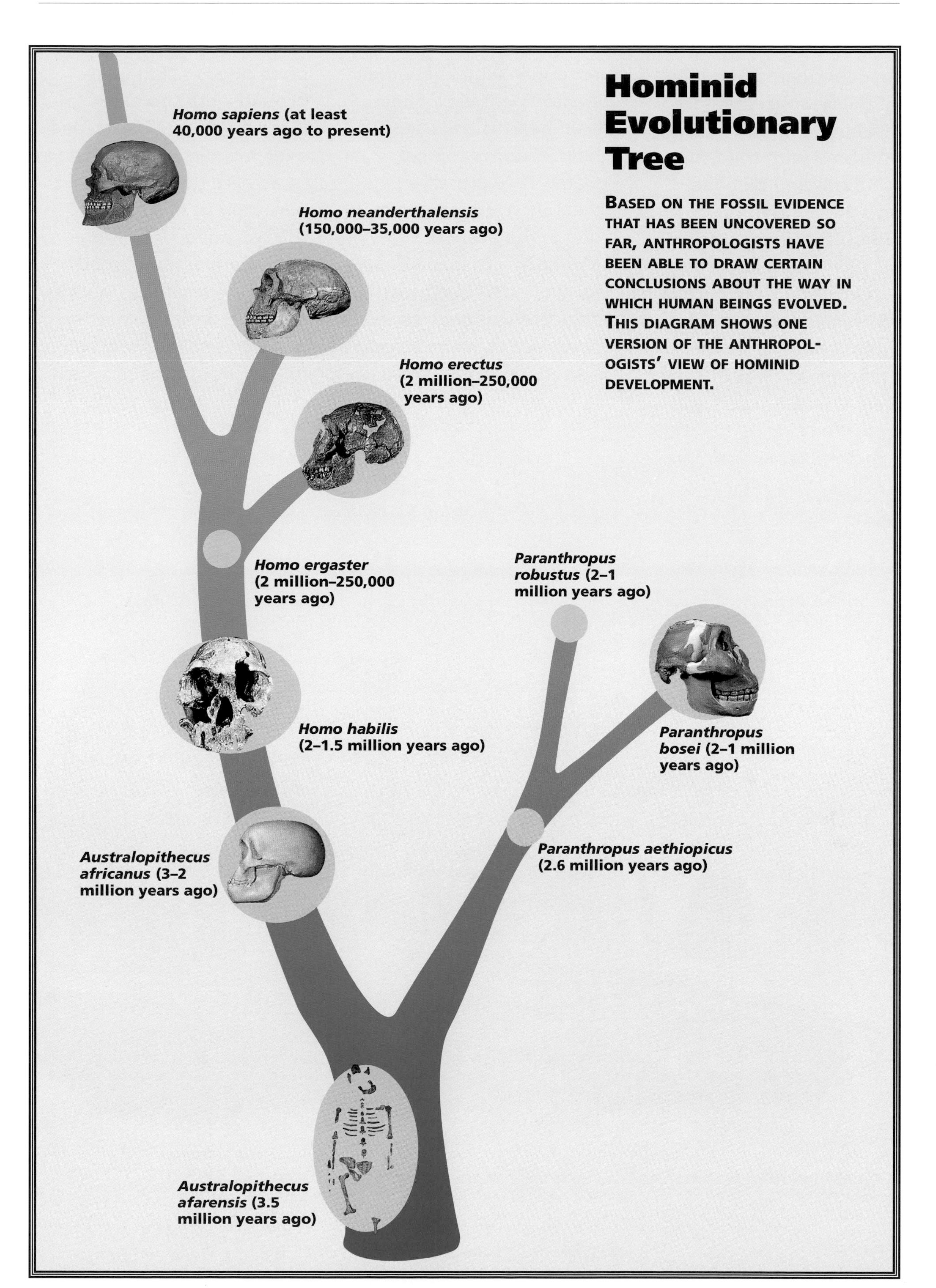
Hominid Evolutionary Tree
BASED ON THE FOSSIL EVIDENCE THAT HAS BEEN UNCOVERED SO FAR, ANTHROPOLOGISTS HAVE BEEN ABLE TO DRAW CERTAIN CONCLUSIONS ABOUT THE WAY IN WHICH HUMAN BEINGS EVOLVED. THIS DIAGRAM SHOWS ONE VERSION OF THE ANTHROPOLOGISTS' VIEW OF HOMINID DEVELOPMENT.
Homo sapiens (at least 40,000 years ago to present)
Homo neanderthalensis (150,000–35,000 years ago)
Homo erectus (2 million–250,000 years ago)
Homo ergaster (2 million–250,000 years ago)
Paranthropus robustus (2–1 million years ago)
Homo habilis (2–1.5 million years ago)
Paranthropus bosei (2–1 million years ago)
Paranthropus aethiopicus (2.6 million years ago)
Australopithecus africanus (3–2 million years ago)
Australopithecus afarensis (3.5 million years ago)

they conquered and killed all the Neanderthals. Or maybe the Neanderthals interbred with the Cro-Magnons and blended into the population. Scientists are not sure.

Upper Paleolithic fossils and artifacts from Europe tell us some remarkable things about *H. sapiens* of that time and place. They were skilled hunters and fishermen who knew their environment well. They built hearths and used heated stones to heat water. They preserved meat by storing it in pits dug in the ground. Cro-Magnons also made beautiful tools, described with the adjective **Aurignacian**, of bone, antler, ivory, wood, and stone. They crafted fancy jewelry and made bone needles that were probably used to make tailored clothing. They traded over vast distances in order to acquire the materials necessary to make these things. Indeed, Cro-Magnons were accomplished artists, who carved figurines of stone, antler, and ivory and painted vivid and colorful pictures (some of which still exist) on cave walls.

A CRO-MAGNON LIVING SITE; CRO-MAGNONS WERE SKILLED HUNTERS AND TALENTED ARTISTS.

Wonderful Walls

In 1985, **Henri Cosquer** dove into the Mediterranean Sea along the coast of France and found one of the greatest discoveries in Paleolithic art. At the base of a cliff under the water, Cosquer found the opening to a cave. When he went inside, he found that the opening led to a huge cave whose walls were decorated with spectacular paintings. These paintings were created by Cro-Magnon peoples 18,440 years ago.

Cosquer Cave (named for its discoverer) is only one of several caves in Europe whose walls are decorated with magnificent artwork by early *Homo sapiens*. A cave in **Lascaux**, France, is well known for having over 1,500 paintings and engravings of deer, bison, horses, and other animals. Another cave, in **Altamira**, Spain, was the first painted cave to be discovered. Found in 1879, it was decorated by Cro-Magnon artists 14,000 years ago. To paint within these pitch-dark caverns, Cro-Magnon peoples used stone lamps that burned animal fat.

PREHISTORIC DEER DECORATE A WALL OF THE FAMOUS CAVE IN LASCAUX, FRANCE.

Getting Closer: The Iceman

After the **Stone Age** (which included the **Paleolithic**, the **Neolithic**, and a middle period called the **Mesolithic**), people in the Old World entered a new period of human development: they began to use metals to make better tools, weapons, and jewelry. With this new technology came new advancements, like writing and arithmetic, and new and useful inventions, like the plow and the wheel.

People first began to use copper around 6,000 years ago in southern Europe and western Asia. This new technology spread, developing at different times in different parts of the world. In time, copper was replaced by bronze, which is harder. Later still, bronze was replaced by the even harder iron. By 2,500 years ago, much of the world had entered the **Iron Age**.

In 1991, an important discovery was made by two German hikers traveling in the Alps. By accident, they came across the mummy of a **Copper Age** man now known as the **Iceman**. The Iceman had died in the extreme cold 5,300 years ago and had stayed in the ice for that long. The cold kept his skin from decaying, so his body was naturally mummified.

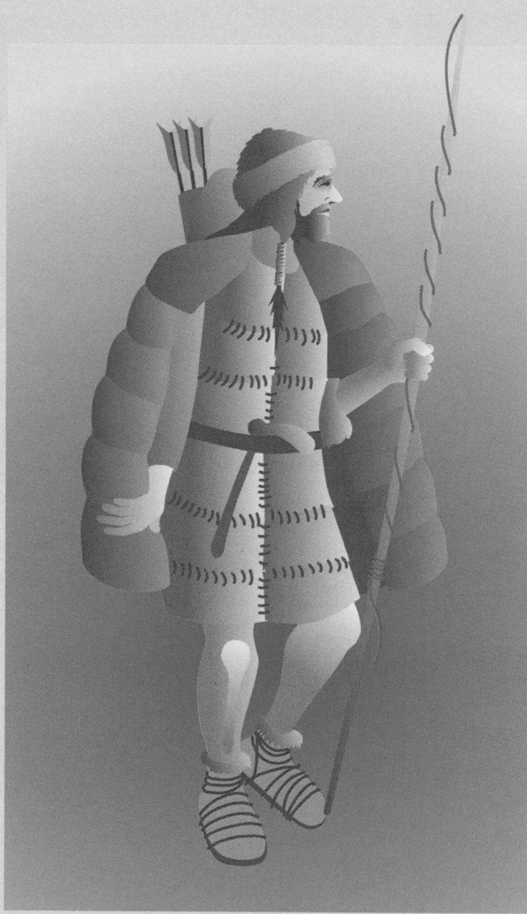

Scientists have been able to tell a lot about the Iceman, such as his age, height, and even that he died in the autumn. But the most important thing about the Iceman's discovery was the collection of tools and other items found with him. Items made of soft materials, such as plant fiber and animal skin, usually decay quickly. But the Iceman's possessions were well-preserved in the snow and ice. These items include one of the oldest copper axes ever found, a finely made flint dagger, rope, pieces of clothing, and a stitched leather bag. Finding these objects has been of great value to researchers. Not only do these artifacts tell part of the story of one man's life, they reveal a great deal about human technology as it existed over 5,000 years ago.

The last phase of the Stone Age was called the Neolithic, and during this period early *Homo sapiens* took primitive tool making to its most sophisticated level. Neolithic tools were made of stone, antler, or bone, among other hard materials. Above, top to bottom: Three-inch (7.6-centimeter) stone knife from the Neolithic period, found in Suffolk, England; Neolithic arrow points from Israel (8,300–4,000 B.C.). Right: This site in South Africa is thought to have been a "stone workshop" for Neolithic tool makers. The chips and flakes that are scattered in a circle around the large stones indicate that stone tools were made in this area.

A WOMAN OF THE NEOLITHIC PERIOD TENDS HER PIGS.

Human beings had become fully modern. But another important event in the evolution of our species was still to come. About 10,000 years ago a great change took place: people began to farm and raise animals. This event is called the **Neolithic Revolution** or the **Great Transition**. It began in southwest Asia and caught on around the world. Before the Neolithic Revolution, people depended mostly on hunting wild animals and gathering wild plants for their food. To do this, they had to move around a lot. But once they began farming and raising animals, they were able to settle down in one place. Villages arose. Populations grew. Because people were living together in villages, they began to need certain things from one another. So some people became craftsmen, some became builders, some became record keepers, and so on. Villages grew into cities, and cities grew into great civilizations around the world.

La Brea Woman

One of the earliest human fossils found in the United States is from a woman who lived about 9,000 years ago. Her fossil bones were found at the **Rancho La Brea Tar Pits** in what is today Los Angeles, California. Known as **La Brea Woman**, she is thought to have died at the hand of another because of the large, clean hole in the side of her skull. Like all hominid skeletons found in the New World, she belongs to the species *Homo sapiens*.

Neolithic Homes

All of the dwellings represented here date from 8,000 to 5,000 B.C. The Neolithic Revolution changed the way in which humans lived, and their homes reflected those changes.

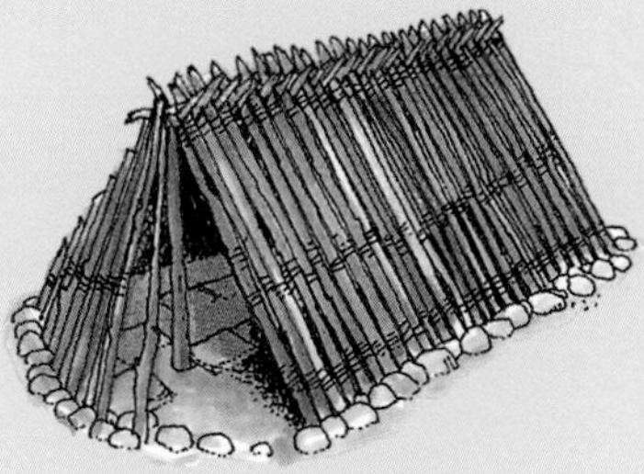

Lepenski Vir, Danube, former Yugoslavia

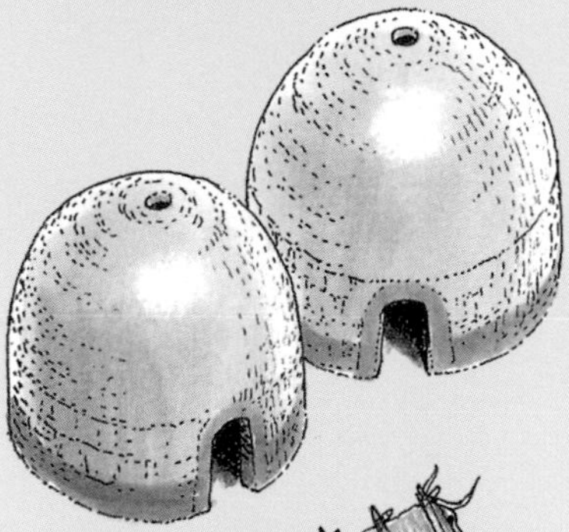

Khirokitia, Cyprus

Karanovo, Bulgaria

Jarmo, Mesopotamia

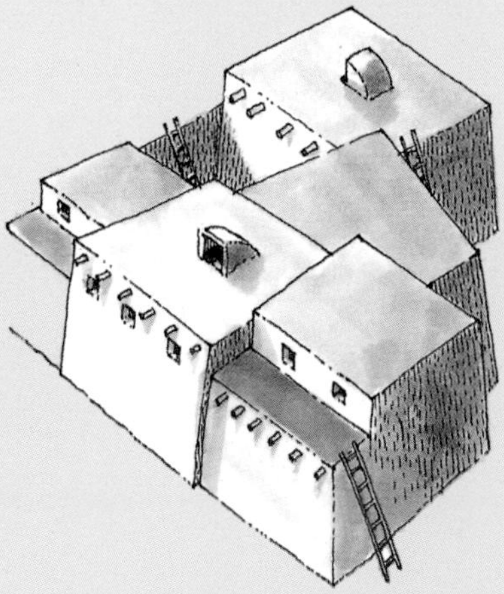

Catal Huyuk, Turkey

The Long Journey

Paleontologists believe that somewhere between 40,000 and 12,000 years ago, *Homo sapiens* began walking into North America. This was during the Ice Ages, when a lot of the earth's water was frozen in the form of glaciers. This left a lot of land exposed that otherwise would have been underwater. At one such place, a natural bridge of land connected the continents of Asia and North America, from Siberia to Alaska. Over time, many animals in search of food crossed over this area, called the **Bering Land Bridge**. These animals were followed by the people who depended on the animals for food. Unaware that they were entering a new continent, they eventually spread all over North America and made their way down to Central and South America. These people were the first Native Americans, known as **Paleoindians**.

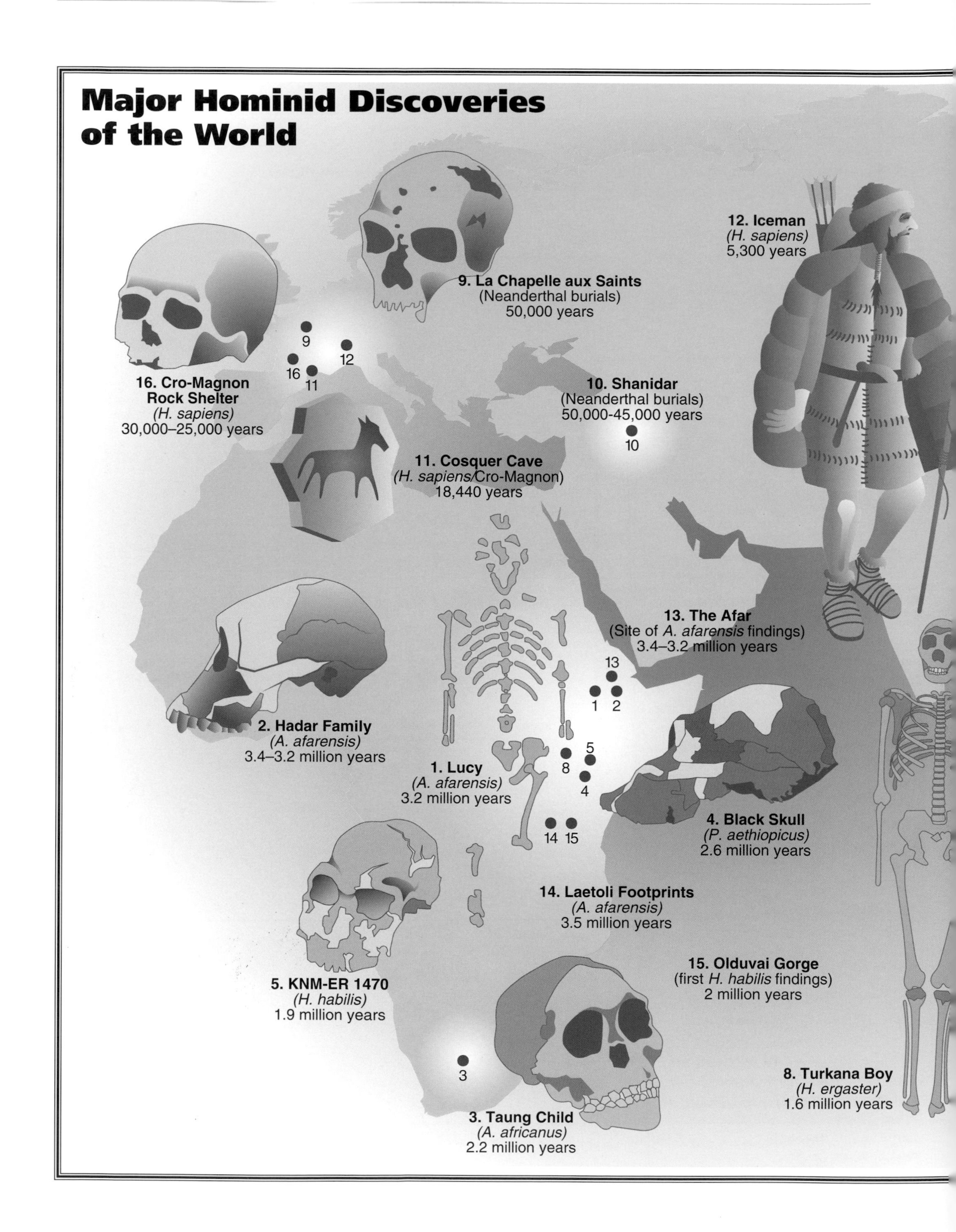
Major Hominid Discoveries of the World
9. La Chapelle aux Saints
(Neanderthal burials)
50,000 years
12. Iceman
(H. sapiens)
5,300 years
16. Cro-Magnon Rock Shelter
(H. sapiens)
30,000–25,000 years
10. Shanidar
(Neanderthal burials)
50,000-45,000 years
11. Cosquer Cave
(H. sapiens/Cro-Magnon)
18,440 years
13. The Afar
(Site of A. afarensis findings)
3.4–3.2 million years
2. Hadar Family
(A. afarensis)
3.4–3.2 million years
1. Lucy
(A. afarensis)
3.2 million years
4. Black Skull
(P. aethiopicus)
2.6 million years
14. Laetoli Footprints
(A. afarensis)
3.5 million years
15. Olduvai Gorge
(first H. habilis findings)
2 million years
5. KNM-ER 1470
(H. habilis)
1.9 million years
8. Turkana Boy
(H. ergaster)
1.6 million years
3. Taung Child
(A. africanus)
2.2 million years

7

7. Peking Man
(H. erectus)
400,000 years

6. Java Man
(H. erectus)
700,000 years

6

This map of the continents of the world shows the locations of some of the most important discoveries made by anthropologists and archaeologists studying the development of early humans. Each location is marked with a number that corresponds to a description of the discovery, including the time period from which the discovery dates. Every one of these discoveries has given us insight into the ways in which humans evolved.

What Lasts Over Time?

After something is buried in the ground, bacteria and chemicals in the soil start to break it down and the object will begin to decay. Some things are more durable than others and will last longer. Generally, things made of soft materials, such as wood, skin, or cloth, will decay quickly. Things made of hard materials, such as stone, bone, or clay, will last much longer underground.

Based on what they find, archaeologists and paleontologists often discover things about what they don't find. For example, a piece of clothing from 20,000 years ago would most likely decay. But a bone needle from that time would be more likely to last. Finding the needle tells the paleontologist that people made clothing, even though the clothing itself has decayed.

Similarly, at some Cro-Magnon sites, paleontologists have found skeletons covered with fancy ornaments made of shell and other hard materials. Through an examination of where the ornaments were placed on the skeletons, paleontologists think the ornaments were at one time attached to clothing, like buttons.

Try the following experiment to get a better idea of what lasts over time.

You will need: a trowel, an area of dirt outside where you can dig a hole, a pencil, paper, and a collection of assorted objects that you will bury.

1. Collect several items made from different kinds of materials. Some of the materials should be soft and others hard. You might include a glass jar, an apple, a metal coin, a sheet of newspaper, a plastic spoon, a chicken bone from your dinner table, and similar items.

2. Make a list of the objects you have collected.

3. Bury the objects in a patch of ground outside your home. (Be sure to select an area where there is little disturbance from people or animals passing by.)

4. Make a map of the spot where your objects are buried so that you can remember where it is.

5. Leave the objects underground for a few months and then dig them up. Examine them to see what has lasted over time and what has not.

THE TWENTIETH CENTURY HAS BEEN MARKED BY SEVERAL OUTSTANDING HUMAN ACHIEVEMENTS, ONE OF THE MOST MEMORABLE BEING THE FIRST MANNED LANDING ON THE MOON'S SURFACE. ON JULY 20, 1969, COPILOT EDWIN "BUZZ" ALDRIN (SEEN HERE) AND PILOT NEIL ARMSTRONG (WHO TOOK THIS PHOTOGRAPH) BECAME THE FIRST TWO HUMAN BEINGS TO WALK ON THE MOON.

THE DIVERSITY OF HUMAN LIFE

Scientists have different theories about the road our evolution took. The further back in time we try to see, the harder it is to figure out which species are our direct ancestors and which species branched off in other directions and became extinct. Knowledge of our ancestry is constantly changing as new evidence is discovered and explored. What we are sure of, though, is that once we became *H. sapiens*, our species really took off in its development. It took nearly 4 million years for *H. sapiens* to appear. Now our technology lets us communicate by computer and send spacecraft out into the reaches of space.

All over the world human beings are alike, from the way our bodies function to the kinds of emotions we feel. All of us need the same basic things to keep us alive and happy, like nourishment, shelter, and love. But just as no two individuals are the same, no two groups of people on earth are exactly alike. Each group has its own customs and its own ways of fulfilling its needs. Is the human species still evolving? Some scientists say that life on earth continues to change, and human beings are no exception. If you look carefully at the human body, you will find certain things there that modern people no longer need. For example, the back molars, or **wisdom teeth**, that show themselves in the late teenage years are no longer useful to modern people. At one time in our evolutionary past, they were probably needed to help our ancestors chew very tough foods. Today, wisdom teeth are not used for chewing (some people never even grow them) and often must be removed by the dentist because they sometimes don't grow in properly. Generations from now, it is probable that no one will have wisdom teeth at all.

Other scientists believe that the human body will change very little in the future, if at all. They argue that, because modern humans have greater control over the environment than our ancestors did, our bodies are less likely to change. It is more likely that we will change our environment to suit our bodies. One way or the other, only paleontologists of the future will know for sure.

A SYMBOL OF HUMANKIND'S REMARKABLE ACCOMPLISHMENTS, THE SPACE SHUTTLE LAUNCHES INTO SPACE.

IN LIMA, PERU, SKULLS TAKEN FROM INCA BURIALS LINE THE STORAGE SHELVES OF THE NATIONAL MUSEUM OF ANTHROPOLOGY AND ARCHAEOLOGY.

BIBLIOGRAPHY

Children

Bruun, Ruth Dowling, and Bertel Bruun. *The Human Body: Your Body and How it Works.* New York: Random House, 1982.

Eldredge, Niles, Douglas Eldredge, and Gregory Eldredge. *The Fossil Factory.* Reading, Mass.: Addison-Wesley Publishing Company, Inc., 1989.

Lasky, Katherine. *Traces of Life: The Origins of Humankind.* New York: Morrow Junior Books, 1989.

Adult

Burenhult, Goran, ed. *The First Humans: Human Origins and History to 10,000 B.C.* San Francisco: Harper, 1993.

______. *People of the Stone Age: Hunter-Gatherers and Early Farmers.* San Francisco: Harper, 1993.

Goodall, Jane. *Through a Window: My Thirty Years with the Chimpanzees of Gombe.* Boston: Houghton, Mifflin and Company, 1990.

Leakey, Richard. *The Origin of Humankind.* New York: Basic Books, 1994.

Milner, Richard. *The Encyclopedia of Evolution: Humanity's Search for Its Origins.* New York: Facts on File, 1990.

Schwartz, Jeffrey H. *What the Bones Tell Us.* New York: Henry Holt and Company, 1993.

Tattersall, Ian. *The Human Odyssey: Four Million Years of Human Evolution.* New York: Prentice Hall General Reference, 1993.

Text Books

Nelson, Harry, and Robert Jurmain. *Introduction to Physical Anthropology.* 3rd ed. St. Paul, Minn.: West Publishing Company, 1985.

Solomon, Eldra, and Gloria A. Phillips. *Understanding Human Anatomy and Physiology.* Philadelphia: W.B. Saunders Company, 1987.

Thomas, David Hurst. *Archeology.* 2nd ed. Fort Worth, Tex.: Holt, Rinehart and Winston, 1989.

Magazine Articles

Laitman, Jeffrey T. "The Anatomy of Human Speech." *Natural History,* August 1984.

Roberts, David. "The Ice Man: Lone Voyager from the Copper Age." *National Geographic,* June 1993.

INDEX

Numerals in italic indicate illustrations.

PHOTO CREDITS

Peter Arnold: © Nelson Max/LLNL: 17t

© Ronald H. Cohn/The Gorilla Foundation: 29

© Bruce Coleman: 16t; © K. & K. Ammann: 27b; © Frank Brown: 36; © E.R. Degginger: 9, 12t, 14t, 35b, 41t, 44, 45t, 45bl, 48r, 55t; © John Fennell: 55; © Jeff Foott: 26; © Bob Gossington: 43r, 45m, 55b; © Linda Koebner: 28b; © Douglas Mazonowicz: 53; © Shaw McCutcheon: 61b; © Patricio Robles: 11t; © L. West: 27t

Dembinsky Photo Associates: © Skip Moody: 15l

FPG International: 10t; © Gary Randall: 28t; © Ken Ross: 17b; © Telegraph Color Library: 21bl, 61t

© Ken Lax: 9, 13, 18, 19, 20, 25, 33, 43, 59 (all photo sequences involving children)

© Los Angeles Museum: 56b

The Waterhouse: © Stephen Frink: 15r

Tom Stack & Associates: © John Cancalosi: 32t, 32b, 33tl, 33bl, 35t, 35m

Tony Stone Images: © Steve McCutcheon: 39t; © Chris Schwartz: 8; © Phil Schermeister: 37

© Visuals Unlimited: 10b, 40l, 49t, 49b; Frank Awbrey: 42r; © Cabisco: 39b; © R. Feldman: 12b; © R. Kessel: 21lr; © National Museum of Kenya: 42l, 46l, 46r; © David Phillips: 20l, 23l

Illustration credits:
© Laura Pardi Duprey: 16, 19l, 21, 22, 23, 24l, 25l, 29, 40, 56, 68

© George Gilliland: 14–15b, 17r, 30, 46, 54, 58

© American Museum of Natural History/New York: 37t

Illustrations by Siena Artworks/London, England, © Michael Friedman Publishing Group 1995: 11r, 24r, 29b, 31, 34, 38, 40r, 47, 48, 52, 56, 57

Key to credits: t=top, b=bottom, l=left, r=right, m=middle